METEORITE

The Earth series traces the historical significance and cultural history of natural phenomena. Written by experts who are passionate about their subject, titles in the series bring together science, art, literature, mythology, religion and popular culture, exploring and explaining the planet we inhabit in new and exciting ways.

Series editor: Daniel Allen

Meteorite

Maria Golia

REAKTION BOOKS

*To John Polk Allen, aka Johnny Dolphin, whose love of Earth
and humankind is present in his every endeavour.*

Published by
Reaktion Books Ltd
33 Great Sutton Street
London EC1V 0DX, UK
www.reaktionbooks.co.uk

First published 2015

Copyright © Maria Golia 2015

Printed and bound in China

A catalogue record for this book is available from the British Library

ISBN 978 1 78023 497 7

CONTENTS

WILLAMETTE

Introduction

And there is a luminous point where all reality is rediscovered,
only changed, transformed . . . And I believe in mental meteorites,
in personal cosmogonies.
Antonin Artaud, *Nerve Scales* (1925)

Like most people, I first saw meteorites in a museum where
they are typically displayed as an extension of mineral exhib-
itions, sometimes in special rooms. These are occasionally called
'cabinets', a word whose original meaning is 'room' and that
evokes 'cabinet of curiosities', those collections of natural and
man-made wonders assembled by Renaissance aristocrats that
were the precursors of museums. The meteorite cabinet of the
National Museum of Prague, before its recent renovation, occupied a corner of that venerable nineteenth-century edifice
beneath a lofty dome, a sequestered space as ornate as the nave
of a church. To get there you had to pass through the mineral
display, a becalming meander among hundreds of pew-like
aisles with glass cases holding specimens of every colour and
texture, each perched on a black oval pedestal, its name engraved
in gold.

After the bemusing pageantry of minerals, entering the
meteorite room elicits a strange excitement. In Prague, one was
greeted by several large, dark masses whose scorched and pitted
surfaces recalled their bombastic journey to earth. The walls were
lined with displays of gleaming slices of iron meteorites mounted
on red velvet like relics or jewels, some as large and shining as
mirrors, each named after the places where they happened to
fall: 'Silver Crown', 'Rodeo', 'Mount Joy'. I photographed my
reflection in 'Xiquipilco', unaware that I'd one day write a book
about meteorites that would offer opportunities for reflections
of other kinds.

Boys in the voids of
the 15.5-ton Willamette
meteorite, on display at
the Hayden Planetarium,
American Museum
of Natural History,
New York, 1911.

But first, what exactly is a meteorite? Although this book is not a technical work, some basic definitions are in order. Meteorites are often confused with meteors. A meteor or 'shooting star' is a small particle or much larger mass of cosmic debris (unhelpfully called 'meteoroids') that makes a visible streak of light as it is incinerated by its passage through Earth's atmosphere. Meteor showers occur when Earth's orbit intersects the streams of meteoroids trailed by comets, 'dirty snowballs' made of rock, dust, ice and frozen gases that orbit the sun. 'Fireball' is another word for bolide, an exceedingly bright meteor that typically burns up in the atmosphere's higher reaches. Depending on their size, speed and composition, some fireballs may explode when striking

Infrared image of Comet Siding Spring, discovered in 2007 by observers in Australia, captured by NASA's Wide-field Infrared Survey Explorer (WISE). The comet appears red because it is more than ten times colder than the surrounding stars, such as the bright-blue one in the foreground.

Vimperk fireball,
31 October 2000.
Digital image taken
with the 'all-sky'
automated camera of
the European Fireball
Network from Ondřejov
Observatory, Czech
Republic.

the atmosphere, producing many fragments. In other cases, part of the meteoroid burns away during flight (a process called 'ablation') but the rest survives the journey to Earth. Only masses that reach the ground are called meteorites.

The oldest things on Earth, meteorites are usually pieces of asteroids, those great lumps of matter left over from the birth of the solar system nearly 4.6 billion years ago that still circle the sun. Asteroids congregating in orbital belts between Mars and Jupiter sometimes crash together, sending random chunks our way. A meteorite may also originate on another planet, ejected into space as a result of a collision with an asteroid. Depending on the size of the mass, its composition and original velocity, it may arrive at speed or explode in the air, scarring the ground with a crater, or else, slowed by the soupy atmosphere, thump down like any object dropped from a height. Meteorites are classified into three main groups because of their predominant mineral compositions: stones (the most common type), irons and stony-irons (the least common). The elliptical stretch of ground over which meteorites tend to fall is called a strewn field. Meteorites are referred to as 'finds' when located some time after they've arrived and 'falls' when their arrival is witnessed and recorded, a relatively infrequent occurrence. 'Meteoritic phenomena' refers to meteors, comets and fireballs as well as to meteorites.

The reader may now be wondering what a book about meteorites is doing in a series of publications about Earth. While it is true that they do not originate on our planet and are indeed among the rarest substances found here, meteorites have indelibly marked humanity's outer and inner worlds. They have acted as geological forces, shaping Earth with their impacts, disrupting the atmosphere and detouring the course of evolution. In the realm of science they have triggered revolutionary reassessments of the solar system, Earth's place in it and the origins of life thereon. As an elemental force arriving from the sky, they figure in myth and folklore worldwide and have likewise insinuated themselves into the consciousness of writers and artists, shedding light on humanity's collective imagination, its intellect and intuition, aspirations and fears. Rather than focusing on the scientific properties of meteorites, this book surveys their pervasive influence on Earth's physical and cultural history, the human response to and interaction with meteorites and the range of sensibilities susceptible to their charms.

With its panoramic camera, NASA's Mars Exploration Rover *Opportunity* photographed this basketball-sized meteorite on Mars, the first ever identified on another planet. Readings from the rover's spectrometers determined its composition of mostly iron and nickel; 6 January 2005.

To illustrate meteorites' timeless, protean attraction, the present work is organized according to the endeavours they have influenced or inspired. Chapter One is an abridged history of the study of meteorites, showing how they challenged the boundaries of the known or knowable, obliging nineteenth-century thinkers to abandon long held theories regarding the solar system, and twentieth-century science to embrace them as a geological and evolutionary force. The compelling spectacle of falling meteorites has prompted people to venerate them as messengers or embodiments of gods. Chapter Two explores myth and folklore, surveying the pervasive role of meteoritic phenomenon in shaping the narratives that brought the divine within human reach. Chapter Three describes the twentieth-century rise of meteorites as collectibles and commodities that, like antiquities, are prized for their rarity, age, origin and aesthetic appeal. Chapter Four is a select anthology of literary references to meteorites, past and present, alongside a gallery of artists who have made them their medium and muse. Chapter Five presents a constellation of meteorite-related events, recounting the peculiar fates of some meteorites whose space odyssey, far from ending on Earth, had only just begun.

Everything about them is engagingly strange but the growing fascination with meteorites transcends their scientific, cultural, artistic or commercial appeal. The attraction derives from a discounted human trait – the capacity for awe – that in a world enamoured of technology and distanced from nature's fierce, sublime reality, meteorites have retained the power to reawaken. I hope that readers will find that sense of wonder somewhere in these pages, and forgive whatever omissions or errors they may contain.

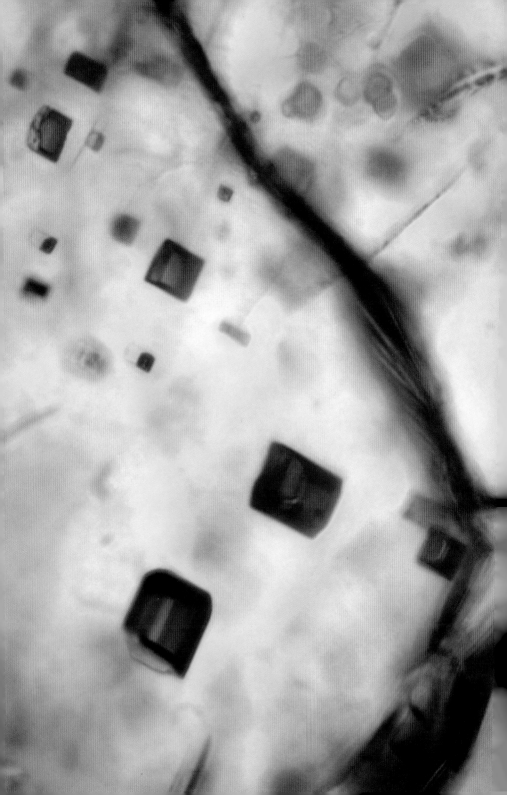

1 Alpha and Omega

On a December afternoon in 1795, seventeen-year-old farm-hand John Shipley was going about his business in the sodden, silent hills of Wold Cottage in Yorkshire when the tranquil scene was shattered by explosions 'like the report of a distant cannon'. Shipley saw something hurtling towards his head that a split second later crashed metres away from his feet, splattering him with stinging shrapnel of cold mud. Two workmates rushed to the side of the dumbfounded young man who, having recovered from his 'extreme Alarm', reported 'that the clouds opened as [the stone] fell and he thought heaven and earth were coming together'.[1]

And so they had. The 25.4-kg mass that had wedged itself into the chalky soil was a piece of an asteroid, by today's classifications a chondrite (stony meteorite), one of the most common types. Among most scientists in Shipley's day the very notion of stones falling from the sky was *infra dig* and eyewitness accounts like Shipley's describing similar events, however consistent in their details, were typically dismissed as the maunderings of credulous peasants. Even Shipley could scarcely believe what he had seen. Later he would note that he had 'ploughed over that very ground last year', as if it had somehow betrayed him.[2]

Angrite meteorite (NWA 4801) photographed at a magnification of 760×. The bright colours emerge under transmitted cross-polarized (pass-through) light.

The history of meteoritics, which began in earnest around the time of the Wold Cottage fall, may be understood as an epic journey into the anomalous. The learned men who first embarked on it subscribed to the Enlightenment ideal of objectively pursuing knowledge but their conventional wisdoms had first to be

The earliest surviving meteorite seen landing in the UK fell at Wold Cottage, Yorkshire. Landowner Edward Topham erected this monument to mark the site in 1799: 'Here, On this Spot, Decr. 13th, 1795 Fell from the Atmosphere AN EXTRA-ORDINARY STONE'. Oil painting by George Nicholson, 1812.

wrenched from their confident grasp. The politics of ideas, religious bias and the difficulty of having to theorize on the basis of second-hand observations conspired to make the reality of falling stones, not to mention their cosmic origin, barely conceivable to the new men of science. Galileo noted that 'the authority of thousands counts for nothing before the single voice speaking the truth', but that depends largely on whose voice it happens to be. Laymen's reports of falling stones were discredited and scientists willing to believe them, obtain samples of the stones or entertain a probable cause for the phenomena were constricted in their thinking by the power of prevailing opinions.

German print dated 1769 depicting the Mauerkirchen meteorite fall of 20 November 1768.

When a fireball exploded over Barbotan in the south of France in 1790, producing a shower of stones, teacher and naturalist Jean F. B. de Saint-Amans asked for official testimonies, expecting none to be forthcoming. He instead received a signed affidavit from a mayor of one of the affected towns, including the depositions of 300 witnesses. Saint-Amans shared it with his friend Pierre Bertholon, editor of the *Journal des Sciences utile* in Montpellier, who published the affidavit not as a testimony to the event but a lament that so many of his countrymen had yet to awaken to the Age of Reason:

> How sad, is it not, to see a whole municipality attempt to certify the truth of folk tales ... the philosophical reader will draw his own conclusions regarding this document, which attests to an apparently false fact, a physically impossible phenomenon.[3]

In a footnote accompanying excerpts of a paper by physics professor Nicolas Baudin, who witnessed the Barbotan fall, the editors of the journal *La Décade* attributed the rocky showers to a kind of mass hysteria:

> The noise that these meteors make in bursting, the dazzling light that they spread, the surprising shock they cause, stuns the majority of those who see them: they do not doubt that the burst had fallen all around them; they run, they look, and if they find, by chance, some little bit of black stone, surely this stone just fell. As the fable spreads, people all over the countryside search for stones and find thousands of them.[4]

For the study of meteorites to begin, their existence had first to be acknowledged, or rather re-acknowledged, since falling stones had been documented throughout antiquity and their possible cosmic origin proposed in the fifth century BC. Science is marked by reversals in thinking and discoveries that defy previous definitions of the natural world. But rather than an advanced theory or mathematical proposition, the birth of meteoritics demanded a return, along a zigzag path, to common sense.

The Royal Society of London, established in 1660 at the avant-garde of the Enlightenment, chose for its motto *Nullius in verba* ('take nobody's word for it') signalling its founders' intention to establish facts via experiments and objective science. In certain matters, however, you had to take somebody's word and who better than Sir Isaac Newton, who served as the Royal Society's president from 1703 until his death in 1727 and whose *Principia* (1687) had expounded the laws of motion and gravity. Newton's orderly universe was the home of 'great bodies, Fixed Stars, planets and Comets' but there were no odd bits flying about. 'To make way for the regular and lasting Motions of the planets and Comets, it is necessary to empty the Heavens of all Matter.'[5] Like other scientists of his time, Newton saw a higher power in nature's workings:

> This most beautiful system of the sun, planets, and comets, could only proceed from the counsel and dominion of an intelligent and powerful Being . . . This Being governs all things, not as the soul of the world, but as Lord over all.[6]

Also like his contemporaries, Newton held ancient thinkers, especially Aristotle (384–322 BC), in deep regard. Aristotle's *Meteorologica*, a treatise on the Earth and its surroundings, maintained that solid bodies other than the sun, moon and planets could not exist in space, a theory congenial to the Christian belief in God's perfect creation held by Newton. According to Aristotle, comets originated between the Earth and moon, and meteors were earth's 'exhalations'. As for meteorites, commenting upon the stone that had fallen at Aegospotami in Thrace (today's Gallipoli Peninsula) in around 469–467 BC, Aristotle claimed it was a terrestrial rock, lifted by strong winds, then fallen back to earth.[7]

The Aristotelian tradition of comprehending the world through observation and searching for 'natural' circumstances through reasoning would eventually ripen into empirical methodologies. But the belief meanwhile lingered that rare events contradicting theoretical models were aberrations, revealing nothing of nature as it 'naturally' functions,

Aristotle depicted in the *Nuremberg Chronicle* (1493).

and therefore unworthy of investigation. As Francis Bacon remarked in *Novum Organum* (1620), 'the human understanding is, of its own nature, prone to suppose the existence of more order and regularity in the world than it finds'. In the late eighteenth century reports of unruly stones were growing too frequent to be dismissed, but few minds were prepared to embrace the degree of disorder required to explain them. William Herschel's discovery of Uranus in 1781 had nonetheless proved that space held some surprises. The planet had been sighted repeatedly over the previous century but was identified as a comet in keeping with accepted knowledge.[8] Some of those willing to entertain the notion that stones did fall maintained, like Aristotle, that they had been somehow thrust into the air, possibly by volcanic activity, yet no one ever saw them going up, only coming down.

The year before the Wold Cottage fall in 1794, a stone that fell at Coscona near Siena compelled Abbé Ambrogio Soldani, a mathematics professor at Siena University, to publish a paper with eyewitness reports and drawings of the stones, concluding they were indeed real and had congealed in the atmosphere. Soldani presented his correspondence with Naples-based mineralogist Guglielmo Thomson, to whom he had sent a sample. Thomson described the stone's black fusion crust and 'quartose' interior speckled with grains of pyrite. The iron component of the stone proved strangely pliant, considering it appeared to have cooled from a molten state and should have therefore been quite brittle. The stone, in short, was different from known terrestrial rocks and Thomson agreed it had formed in the 'high fiery cloud' that was seen approaching Coscona when it fell.[9]

Aristotle's theory that all things on Earth originate with it or in its atmosphere accommodated the possibility that falling stones were 'exhalations' of gas and dust that had somehow solidified. In *On the Nature of Things*, the Roman poet and philosopher Lucretius (*fl.* first century BC) expressed ideas about Earth's atmosphere and the likelihood of masses forming there that had long been accepted:

For whatever ebbs from things is all borne always into the
great sea of air; and unless it in return were to give back
bodies to things and to recruit them as they ebb, all things
ere now would have been dissolved and changed into air.[10]

The Persian polymath Avicenna (980–1037) described falls of
stones and irons that he believed were formed in the atmosphere.
René Descartes (1596–1650) likewise argued that lightning
can make atmospheric dust turn to stone. Benjamin Franklin's
demonstrations of lightning bolts conducting electricity (published in 1754) lent credence to this idea. Antoine-Laurent de
Lavoisier, author of the seminal *Elementary Treatise on Chemistry*
(1789), endorsed the theory that falling stones were formed
by lightning in Earth's atmosphere, as did other influential
authorities.[11]

Abbé Andreas Xaver Stütz was the assistant director of the
Imperial Natural History Collection at Vienna. Stütz's paper of
1790, 'On Some Stones Allegedly Fallen from Heaven', set out
to debunk the notion of a 'heavenly' origin based on reports and
samples of three falls (recorded in 1751, 1753 and 1785).

it was said that the iron fell from heaven. It may have been
possible for even the most enlightened minds in Germany
to have believed such things in 1751 due to the terrible
ignorance then prevailing of natural history and practical
physics; but in our time it would be unpardonable to regard
such fairy tales as likely.[12]

Recent experiments showing how electricity transforms the
chemical compound iron oxide into metal seemed to support
his (and Lavoisier's) conclusion that lightning turned elements
rising from the Earth into stones that fall to the ground. Stütz's
contentions were aligned with Lavoisier's in another way.
Lavoisier was known for dismantling questionable yet popular
beliefs of his day, including water dowsing and the 'animal
magnetism' proposed by Franz Anton Mesmer.[13] The urge to
distinguish science from superstition and to curb public credulity

Mezzotint of the 1783 fireball seen from Winthorpe, Nottinghamshire, signed and captioned by Henry Robinson: 'This extraordinary Phœnomenon was of that Species of Meteor which the great Phisiologist Dr Woodward and others call the *Draco volans* or Flying Dragon.'

was strong, yet many of the 'fairy tales' that were alive in the popular consciousness had their basis in the same ancient corpus that influenced scientific thought.

In 1790, Charles Darwin's grandfather Erasmus Darwin (1731–1802), a physician and natural philosopher, published his poem 'Botanic Garden'

> to inlist [sic] Imagination under the banner of Science; and to lead her votaries from the looser analogies, which dress out the imagery of poetry, to the stricter ones which form the ratiocination of philosophy.[14]

Darwin's lyrical descriptions of various natural phenomena included the genesis of mushrooms, which were believed to sprout where lightning struck. 'So from dark clouds the playful lightning springs, rives the firm oak or prints the fairy ring', he wrote, referring to the arc or circular formation in which mushrooms grow. The notion that mushrooms were spawned by lightning, or thunder as the ancient Greeks and Romans held, was echoed in folklore linking them with meteors, another sudden, mysterious phenomenon.[15]

In a letter to Sir Hans Sloane of the Royal Society in 1733, a Mr Crocker wrote that following the daytime sighting of a bright meteor, he searched for the place where he thought it had fallen, expecting to find 'some of those jellies which are supposed to owe their beginnings to such meteors'. He failed in his search for what was probably *Tremella mesenterica*, also known as 'yellow brain' fungus, described in a British folklore society publication:

> A substance occasionally found after rain on rotten wood or fallen timber, in consistency and colour it is much like genuine butter. It is a yellow gelatinous matter, supposed by country people to fall from the clouds. Hence its second popular name of 'star-jelly'.[16]

Elsewhere it was called 'star shoot', according to the Lincolnshire Folkloric Society, 'a gelatinous substance often found in the fields after a rain, and vulgarly supposed to be the remains of a meteor shot from the stars'.[17] To eighteenth-century advocates of a more rigorous science, it must have seemed that the imagination, especially that of untrained observers, had been enlisted overmuch.

Old ideas had to be questioned, if only because they had gone unchallenged for so long. Pliny, in his *Natural History* (c. AD 77), wrote: 'That Stones fall often, no Man will make any doubt.'[18] And for centuries, few men did. But he also reported rains of milk, blood, flesh, iron, wool, tiles and bricks, phenomena that at the dawn of the enlightened nineteenth century seemed outlandish.[19] Historian Eusebius Salverte summed up the situation in 1803, when an accumulation of knowledge was nonetheless pointing back to the reality of falling stones and forward to proofs of their cosmic origin:

Alex Kuno, *Lil' Meteor*, 2012, acrylic and ink on wood, from the series *Little Tragedies*.

The ancient historians all make frequent mention of the productions of stones [fallen from the atmosphere]. No doubt was maintained respecting them in the Middle Ages; but the difficulty of accounting for them induced us not only to suspend our belief until called forth by more regular observation, which was very prudent, but also, which was less reasonable, to carry with us in this research a predetermination to see nothing, or to deny what we had seen.[20]

Science belonged to the aristocracy and the Vatican counted many a noble scholar and educator in its clerical ranks. The predetermination Salverte mentions was social as well as intellectual. Accepting that stones fell from heaven demanded a reassessment of 'vulgar' beliefs along with the classical Greek and Latin sources embedded in them. The elite had to loosen its monopoly on knowledge and entertain unorthodox ways of viewing a universe that refused to conform to cherished norms. Francis Bacon (1561–1626) formulated the objectivist's dilemma at the dawn of the scientific revolution: 'For what a man had rather be true he more readily believes. Numberless . . . are the ways, and sometimes imperceptible, in which the affectations colour and infect the understanding.'[21]

Largest meteorite recovered from an observed shower of 18 July 2011 in Tata Province, Morocco; 12.8 × 11.3 × 9.7 cm, 1.282 kg.

Unhindered by Aristotelian theories and biblical cosmologies, the Chinese were more conversant with astral phenomenon. While European astronomers adopted the Greek system of constellations, eventually mapping 88, the Chinese divided the sky into 283 constellations. These smaller divisions allowed for a degree of accuracy in identifying the locations of cosmic events only appreciated in the West in the nineteenth century, when the long-term motion of Halley's Comet was determined using Chinese observations of its returns beginning in 230 BC. References to 'guest stars' likewise helped identify the positions of past supernova explosions. By 28 BC Chinese astronomers were regularly recording sunspot observations made with the naked eye.[22] Such reports are scarce in classical sources, owing partly to the belief that the sun had no unsightly blemishes. Sunspots were not studied seriously in Europe until the early seventeenth century, when astronomers (including Galileo) who had observed them with telescopes were still arguing over who had discovered them first.

Astronomical events were recorded in the official histories of Chinese dynasties, assembled by state officials and local scholars of every province and district, who apparently accepted that stones fall from the sky and possibly the firmament. The Chinese phrase for such events was either 'a stone fell' or more typically *hsing-yun*, 'a star fell'. The earliest reference to a meteorite is from 645 BC: 'Five stones fell in Sung.' The following account of an exploding fireball and meteorite fall dates to AD 1064:

> In the morning at Chang-chou, there was a loud noise like thunder from the sky. A large star almost like the Moon was seen at the south-east . . . following another sound of thunder, the star fell into Mrs Hsu's garden . . . It was seen by everyone near and far. Its flames brightly lit the sky.[23]

Similar passages appeared in a chapter of the histories devoted to 'ominous and unusual events' that were mostly compiled from reports by average citizens.

The strange sounds witnesses described as accompanying the appearance of large fireballs have yet to be entirely explained. In the eighteenth century, like meteorite falls, the sounds were dismissed as flights of fancy. Cultural factors clearly influenced the way people interpreted the auditory experience. Gregory of Tours in AD 580 recorded 'a sound of as many trees crashing to the ground'; Chinese annals (AD 817) note 'a noise like a flock of cranes in flight', whereas witnesses of the Peekskill, New York, fireball of 1992 heard something 'like a sparkler'.[24] Reports from different times and places mention 'swishing', 'rushing' 'popping', 'buzzing', 'crackling' or 'vibrating' in the air. Aside from implying sensations beyond the auditory, these sounds differ from the explosions and/or sonic booms typically heard seconds to minutes after the fireball is extinguished, in that they occur while it is still speeding across the sky.[25]

Eighteenth-century scientists knew that sound ordinarily took time to travel, as did anyone who had observed the delay between the flash of a distant cannon and its boom. Edmund Halley (1656–1742), who used his friend Isaac Newton's theory of gravitation to determine the orbits of comets, was intrigued by reports of a fireball over England in 1719 and reports of its hissing 'as if it had been very near at hand'. Halley calculated the probable distance of the fireball from where the sounds were heard ('60 English miles') and dismissed the hissing as the 'effect of pure fantasy'. Thomas Blagdon, secretary of the Royal Society, likewise concluded in 1784 that the sounds were the product of 'an affrighted imagination'.[26]

Well into the twentieth century, the sounds were considered 'an observational illusion' owing to the disorientation of people witnessing the sudden appearance of a large, exceedingly bright, preternaturally fast-flying object. 'The explanation is without a doubt psychological', wrote a professor of mathematics and astronomy in 1939.[27] One of the few to object was American meteoriticist H. H. Nininger (1887–1986), who thought the sounds should be regarded 'as a problem in physics rather than psychology' and proposed their 'ethereal as well as aerial propagation' in 1934.[28] The sounds, now described as electrophonic (resulting from the direct conversion of electromagnetic radiation into audible sound), are not yet fully understood. But the enduring view that they were imaginary demonstrates the difficulty of objectively assessing someone else's experience, and the (subjective) urge to dismiss the unknown as impossible.[29] This hyper-wariness, combined with the relative rarity of witnessed fireballs and meteorite falls, helps explain why the science of meteoritics had such a hard time getting off the ground.

By the early 1800s a controversial book, a chemical analysis, some startling astronomical observations and the auspicious coincidence of several well-documented falls combined to shift academic opinion away from conventional disbelief and towards the radical truth. Ernst Chladni's *On the Origin of Ironmasses* (1794) was inspired by a conversation with Georg Christoph Lichtenberg, renowned Göttingen physicist and natural philosopher who understood falling stones as 'cosmic phenomena'. Chladni spent the next three weeks in the Göttingen University Library, studying detailed reports of 24 fireballs and eighteen stone falls, alongside less carefully documented cases, including ten historical falls dating from the first to seventeenth centuries.[30] He was surely intrigued by the mentions of strange sounds. Now referred to as 'the father of acoustics', Chladni found a way to make sound waves visible in 1787 by drawing a violin bow along the edge of a metal plate scattered with fine sand, which separated into elegant, complex patterns.[31] Although his first loves were music and mathematics, Chladni had studied law, and in the testimonials of fallen stones to which the scientific community had turned a deaf ear, he discerned the ring of truth.

Chladni collated similarities in the reports, concluding that falling stones were linked to fireballs and theorizing that the latter had a solid core that grew incandescent from friction with Earth's atmosphere. These masses, he maintained, were cosmic debris that had failed to form into planets or else were created by explosions or collisions between planetary bodies.[32] Having examined reports of the stones' chemical composition and noting iron and other familiar elements, he correctly postulated that all planets are made of the same essential stuff. It seems possible that Chladni or his mentor Lichtenberg were aware of Immanuel Kant's *Universal Natural History and Theory of the Heavens* (1755). In his nebular hypothesis Kant deduced that the solar system began as a spinning disc of gases that coalesced into planets, shedding matter in the process, an idea that may have influenced Chladni's understanding of meteorites.[33] In any case, he insisted on their cosmic origin and rejected all other explanations for the falling stones: atmospheric, volcanic ejecta from Earth or, as some proposed, from the moon. Anticipating hostility, Chladni pointed out that Newton's empty space was no less 'preposterous' than his more cluttered one, since 'neither one can be proved on *a priori* grounds'.[34] Some colleagues agreed with him in private but refused to do so publicly. Although Chladni is now acknow-ledged as 'the father of meteoritics', it would take 150 years for several of his daring theories to be validated.

Chladni's book reached England by 1796, the same year the Wold Cottage stone was displayed in Piccadilly Circus, where a curious public paid a shilling each to view it. Sir Joseph Banks, president of the Royal Society, who possessed specimens from Wold Cottage and other fallen stones, was compelled to com-mission a comparative chemical analysis. Edward C. Howard and Jacques-Louis, comte de Bournon, studied stones from sev-eral witnessed falls in addition to samples of mysterious so-called 'native irons', including the Table of Iron from Argentina and the Pallas Iron from Siberia. The chemists identified common components, including 'a coating of black oxide of iron' (fusion crust), 'curious globules' (silicate inclusions called chondrules) and a high nickel content in the iron, all of which distinguished their samples from terrestrial rocks. The Royal Society adjusted the title of Howard's paper before publication, betraying a wish to hedge its bets about the nature of the stones. Howard's title, 'Observations on Certain Stony and Metalline Substances which Have Fallen at Different Times on the Earth' became 'Substances which Are Said to Have Fallen'. Howard concluded his paper by asking: 'Have not all fallen stones, and what are called native irons, the same origin? . . . are all or any, the produce of the bodies of meteors?'[35]

These questions, echoing Chladni's assertions, were the more pressing with the first discoveries of asteroids in 1801 and 1802, small bodies that were neither comets, stars nor planets and therefore should not have existed.[36] Then on 26 April 1803, out of a clear blue sky, thousands of stones pelted the French commune of L'Aigle, 142 km from Paris. France's Minister of the Interior dispatched Jean-Baptiste Biot, a young and recently elected member of L'Institut National, to investigate. The first scientist to venture out of the lab or library to assess the phenomenon, Biot was thorough: interviewing eyewitnesses, collecting samples and carefully mapping the stones' strewn field, the elliptical area over which they fell. His engagingly written report of 'without a doubt the most astonishing phenomenon ever observed by man' swayed influential minds and opinions.[37] Although some were willing to accept that the stones may have come from space, the beliefs that they were formed in Earth's atmosphere or produced by lunar volcanoes persisted. Nonetheless, meteorites

The giant asteroid Vesta, discovered by Heinrich Wilhelm Olbers in 1807. Photographed by NASA's *Dawn* spacecraft on 17 July 2011 at a distance of about 15,000 km.

27

One of the thousands of meteorites that pelted L'Aigle, France, on 26 April 1803. This specimen was 5.4 × 2.4 × 2.8 cm and weighed 74.4 g.

had finally passed from fable to fact. The 1824 edition of the French *Dictionary of Natural History* boasted a 46-page entry for *pierres météoritiques* (meteoric stones), marvelling at how 'there was [recently] even some sort of stubbornness from savants to support the refutation (of meteoritic stones) and to ridicule those who were defending their existence'.[38]

The very act of considering meteorites' origins had meanwhile enlarged the cosmic frame of reference for both scientists and the greater public. A spectacular showing of the annual Leonid meteor shower in November 1833, 'a blizzard of light' visible throughout North America, captured the world's imagination.[39] Meteoritic phenomena had traditionally been cast as ominous and those of a religious bent could not resist seeing the shower as the fulfilment of biblical prophecy in the Book of Revelation (6:13): 'and the stars of heaven fell to the earth'. But a growing awareness among the educated public of astronomical discoveries allowed meteoritic phenomena to be viewed with a more self-confident awe. A North Carolina doctor who witnessed the shower of 1833 while making his pre-dawn rounds described his experience:

> The display was in the highest degree magnificent and imposing . . . every region of the atmosphere all the while presenting the sublime spectacle of a shower of fire. I was

Leonid meteor shower
of 1833 seen from
Niagara Falls, from
Edmund Weiss,
*Bilderatlas der
Sternenwelt* (1888).

startled by the splendid light . . . rendering even small objects
quite visible; but I heard no noise, though every sense seemed
to be suddenly aroused, if I may so speak, in sympathy with
the violent impression of the sight.[40]

Another account, from a witness in Missouri, likewise signals
an enthusiastic curiosity:

We were awakened and told that the stars were falling
and flying in all directions of heavens; and knowing that
the individual who awakened us was a person of observation and science, we instantly hurried from our room . . .
From a vantage point above a prairie, the horizon wide
open afforded an excellent opportunity of beholding
this Phenomenon in all its various aspects, and impressive
sublimity. The most perfect master of language would
fail of conveying to others a full picture . . . and vain
would be his attempt to express the sensations of
its beholders.[41]

Astronomers began scouring Arab, European and Chinese
chronicles for references to the Leonids, tracing their recur-
rences back through antiquity and calculating an approximately
33-year interval between 'peak' showings such as that of 1833.

Meteoritic phenomenon inspired this pooling of data across
time, place and space, heralding a more globalized scientific
endeavour. As for the public, stargazing became so popular a
hobby in England that it was lampooned by the editors of *Punch*
in an article entitled 'Meteors for the Millions' (1861).[42] In
Russia, the curator of the Moscow Museum's meteorite collection, Vladimir I. Vernadsky (1863–1945), sent word across the
steppes, soliciting reports of falls and finds from the citizenry.
The study of meteorites contributed to Vernadsky's vast intellectual repertoire, later synthesized in a landmark generalized
concept of the natural sciences, *The Biosphere* (1926), his treatise
on interactive planetary and cosmic processes.[43]

In 1858 Karl Reichenbach of the Prussian Academy of
Science presciently wrote that a meteorite is 'a cosmological,
astronomical, physical, geological, chemical, mineralogical and
meteorological object', and valuable to all these studies. But in
the 1920s meteoritics was still 'a minor pursuit . . . not quite
respectable' and historians of science ignored the import of
Chladni's work as late as 1970. In what must surely rank as
one of the history of science's greatest reversals, meteoritics
have since passed from 'a narrow specialty' to a cornerstone of

The Leviathan of
Parsonstown. Built in
1845 by Irish nobleman
William Parsons, the
third Earl of Rosse, in
County Offaly, Ireland,
the telescope had
a mirror reflector
weighing 4 tons and
a 16.4-m-long tube.

planetary studies.[44] Those dubious falling stones transformed
Earth's place in the universe by tethering it to the stars.

Analyses of meteorites' chemical and mineral components
continue to illuminate Earth's interactions with space, the age
and composition of our planet and the solar system and the
evolution of stars.[45] Meteorites have provided us with samples,
free of charge, of the moon and Mars. Some carry stardust,
trapped before the birth of our solar system, a time, wrote
Lucretius in around 55 BC, of 'strange stormy crisis and medley,
gathered together out of first beginnings of every kind'.[46]
Meteorites are not only a virtual goldmine of data, but recent
studies seem to indicate that Earth's reserves of gold are themselves the result of massive meteorite bombardments.[47] Billions
of tons of asteroidal material were stirred into the planet's mantle 200 million years ago and by 'gigantic convection processes'
may have concentrated precious metals in the ore deposits mined
from antiquity till now. A scientist involved in the research
called the calamitous creation of humanity's economic exchange
a 'lucky coincidence'.[48]

Meteorite impacts have helped shape Earth's physical features. A crater near Manson, Iowa, now flattened into the plain
by 2.5 million years of glacial erosion, was nearly 5 km deep
and 32 km across, produced by a 10-billion-ton rock moving

200 times faster than the speed of sound.[49] Aside from gouging Earth's crust, some impacts released enough energy to steam dozens of metres off the top of the planet's seas. Ongoing research suggests that meteorites may have meanwhile delivered signature ingredients to the primal soup. Some carry traces of the amino acids that build proteins featuring both in organic structures and in enzymes, the catalysts that accelerate or regulate chemical re-actions.[50] Intimately implicated in the nature of life on Earth, meteorites have served as signposts on the winding path of humanity's self-knowledge. But just as they have helped author our planet's epic from its beginning, contributing geological and evolutionary plot twists, so too may meteorites write the story's end, as agents of annihilation. What space giveth, space also taketh away.

Complete slice of the Shis'r Lunar meteorite, found by meteorite hunter Michael Farmer in Dhofar, Oman, 4 January 2008; 5.5 × 3.8 × 0.1 cm, 3.32 g.

The solar system has filled up a lot since the dawn of meteoritics. Jupiter, which had four known moons in the eighteenth century, now has 50.[51] Four asteroids were known by 1807; as of 2013, over 600,000 were catalogued but by some estimates they number in the millions, some little more than flying rocks, the biggest one-third the size of Earth's moon.[52] Typically depicted as orbiting in a thickly peppered belt between Mars and Jupiter, some 200 million km from Earth, the space between asteroids is immense, though not as great as the space between planets. Agoraphobics are advised not to contemplate what the solar system looks like drawn to scale. Reduce Earth to the size of a pea and Jupiter would be 300 m away, Pluto 2 km away 'and about the size of a bacterium'. The closest star, Proxima

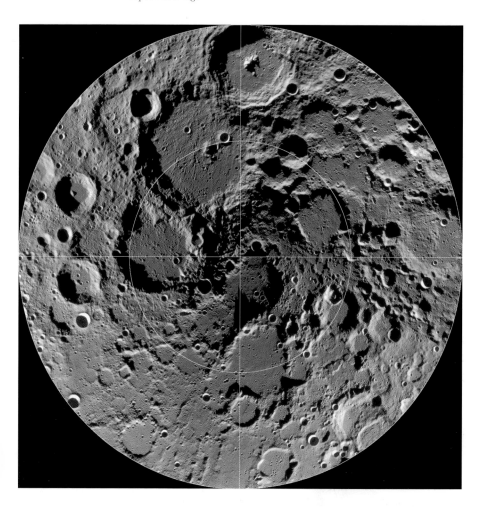

Unlike Earth, the moon has no atmosphere to shield it from impacts. Its battered north pole was captured by NASA's Lunar Reconnaissance Orbiter camera.

Centauri, would be 16,000 km away from the pea-sized Earth.[53] Having calculated the possible number of planets that may exist beyond our solar system, Carl Sagan noted that the emptiness between them is so great 'that if you were randomly inserted into the universe' the chances of landing on or near one would be less than one in a billion trillion.[54] Advances in astronomy suggest that stars and galaxies comprise only 4 to 5 per cent of a universe composed largely of invisible 'dark matter' and the 'dark energies' that drive its expansion.[55] Science may sketch the contours of this reality, but the intellect, even aided by intuition, can neither grasp nor convey the spaciousness of space.

That meteorites should land on Earth may seem extraordinary, but our planet whizzes through the quasi-void around the sun at a speed of 30 km per second, in constant collision with asteroidal, cometary and planetary debris. An impact event with the energy of the Hiroshima nuclear bomb occurs around once a year, while a megaton event is expected at least once a century. No one would be around to calculate these probabilities were it not for Earth's superbly protective atmosphere; most bodies break up when they encounter it and are fragmented and dispersed by aerodynamic stresses before they reach its lower realms.[56] More discreet is the microscopic powdering of interplanetary and interstellar dust Earth receives annually by the tens of thousands of tons. Amid it are enough micrometeorites (also known as magnetic spherules, average size 0.2 mm) for at least one to fall on each metre of the planet's surface per year. Every mundane stretch of highway and every makeshift roof in our great cities' slums hosts some tiny relic of the early solar system.[57]

The flux of meteorites is not constant. In its youth and at various points in its 4.5-billion-year history, our planet was severely battered. Geologist Derek Ager compared Earth's life to that of a soldier: 'long periods of boredom and short periods of terror'.[58] Meteorite impacts and the terrific craters some caused were not always recognized as forces in Earth's geological development. At the end of the nineteenth century, geologists and astronomers were looking beyond Earth (where craters were largely unexplored) to the moon, whose pockmarked surface was attributed to volcanic activity. Based on telescopic observations, Grove Karl Gilbert (1843–1918), chief geologist of the U.S. Geological Survey, proposed they were instead formed by impacts. Gilbert was present at the gathering of the American Association for the Advancement of Science in 1891 when mineral merchant Albert E. Foote described the unusual meteorites he found in the vicinity of Coon Butte, Arizona, containing microscopic diamonds. Foote was so intent on announcing his exceptional find that he barely mentioned that Coon Butte was the rim of a gigantic crater, soon to become the first of its kind ever studied on Earth. An estimated 200 million tons of rocks were displaced to form its perfect bowl, measuring nearly 2 km across and 180 m deep. In keeping with prevailing views, Foote thought the crater was volcanic and dismissed the small meteorites lying in its proximity as coincidence.[59]

Gilbert instead proposed that the crater, like those on the moon, was made by an impact and the large mass that caused it might still be buried there. Gilbert soon set out 'to hunt a star', spending weeks examining the crater and hurling mud balls into mud surfaces to observe the cavities they made, concluding that an impact object, which would arrive obliquely, could still leave a round, not elliptical, crater.[60] His estimates of the probable size of the mass (based on the amount of material thrown up to form the crater's rim) were less satisfying. Gilbert's calculations failed to take the crater's 50,000 years of erosion into account; he arrived at a mass the same size as the hole, which made no sense. Nor did he find the magnetic anomalies that a large iron-rich body would have produced. Dismayed but faithful to science, Gilbert was obliged to conclude that the crater was volcanic and that the meteorites around it were coincidental.[61]

To Daniel Moreau Barringer (1860–1929), a quintessential American entrepreneur, this was 'highfalutin' nonsense. Born in Raleigh, North Carolina, Barringer was the son of a u.s. congressman who had shared an office with Abraham Lincoln. At military school, Barringer's disrespect for authority won him an early dismissal. At fifteen he enrolled at the College of New Jersey (later Princeton University), where he befriended classmates who would influence his own and America's destinies, including future u.s. president Woodrow Wilson. Although he graduated in law at the University of Pennsylvania, Barringer as he now called himself, opted out, partnering with a geologist and spending several years prospecting for minerals in Spain, South American, Mexico, California and the southwest United States. In 1897 he hit the jackpot, acquiring the mining rights for the Commonwealth Silver Mine in Pearce, Arizona. He

Stan Gaz, *Meteor Crater
Arizona*, dia. 1.2 km,
49,000 years old. Solarized
black-and-white print,
2009.

married and celebrated his newfound wealth on a yearlong honeymoon around the world.[62]

Barringer heard of the Arizona crater in 1902 at a chance encounter at the Tucson Opera with Samuel J. Holsinger, who had also stepped out for a smoke. A U.S. Land Office agent who knew the territory, Holsinger said that according to local legend, a meteorite formed the bowl. Barringer was intrigued but incredulous, '[because] I realized the crater must have been examined by members of the United States Geological Survey while making topographic maps of the region, and in their report they evidently did not accept this theory'. Holsinger sent samples of the meteorites from around the crater; analysis revealed 92 per cent iron, 5 per cent nickel and trace elements including iridium, platinum and microdiamonds. The metals contained in these samples were valued at U.S.$125 per ton at the time, and Barringer, riding a long lucky streak, figured there were at least tens of thousands of tons.[63]

Unlike Gilbert, Barringer was convinced the meteorite existed and was ready to dig to the antipodes to find it. Apart from the monetary benefits, he was compelled to prove academia wrong about the crater's origin, calling their volcanic and, later, 'steam explosion' theories 'blind' and 'demented'.[64] Moving in the same circles of Theodore Roosevelt and the trustees of Princeton

Stock share for Daniel Moreau Barringer's Crater Mining Company, founded in 1903.

and MIT, Barringer found many enthusiastic investors but they abandoned him as results failed to materialize. Barringer, who took his fortune from the earth, put it back where he found it, drilling unsuccessfully for 26 years to find the meteoritic mother lode. In the words of his grandson, he 'was not just a man of strong opinions, he was a man of constant opinions'. His stubborn prospecting eventually ruined him. Barringer never imagined that the mass (later estimated at 300,000 tons) had vaporized in the bombastic energies of impact (equivalent to approximately nine megatons of TNT) and the small scattered meteorites were all that remained.[65]

A nine-page spread in *National Geographic* magazine titled 'The Mysterious Tomb of a Giant Meteorite' (1928), outlined mining operations at the site without mentioning Barringer or his quest. The article's urbane author described a 'splash in stone' so deep 'that [Lower Manhattan] could be dropped into the hole and only the Woolworth Tower would project prominent above the rim'.[66] The article highlighted Gilbert's investigations, perhaps as a nod to the esteemed Geological Society, even though Gilbert had held firm to the crater's volcanic origin until death. Barringer died in 1929, before the substantial evidence he had gathered would help prove the mandarins wrong.[67]

In the 1950s geologist Eugene Shoemaker brought a unique perspective on Meteor Crater (as it had become known), as he had recently studied nuclear blast zones at a testing field in Nevada and saw parallel signs of destruction. With a colleague he identified coesite, a quartz formed only under intense shock at high temperatures, in samples taken from both the blast zones and Meteor Crater, proving its impact origin conclusively in 1963. Shoemaker's next move was not to seek out other craters, but to look to space for what caused them.[68] In the 1970s, with his wife Carolyn and other colleagues, Shoemaker conducted a systematic survey of the inner solar system, tracking asteroids. Only twelve had so far been discovered, not because science lacked the means, but for want of interest. 'Astronomers in the twentieth century essentially abandoned the solar system', Shoemaker explained in an interview: 'their attention was turned to the stars, the galaxies.' When Shoemaker's team focused the Palomar Observatory telescope closer to home, the findings were startling: asteroids and their detritus crossed Earth's path regularly.[69]

In his *Nature of the Stratigraphical Record* (1973) Derek Ager pointed out that what humans consider rare catastrophes are a routine aspect of Earth's geological process. 'The hurricane, flood or the tsunami may do more in an hour or a day that the ordinary processes of nature have achieved in a thousand years', he wrote.[70] While Ager refrained from naming meteorite impacts as geological game-changers, he cited respected palaeontologist Digby McLaren, who in 1970 had linked them to mass extinctions. At the time, impact-related topics were treated by the scientific community like some private indiscretion: they were admitted but rarely discussed. Geologists and palaeontologists, among others, followed the maxim that 'nature does not make jumps'. Nothing significant occurs owing to some anomalous 'paroxysm', only well-observed, familiar processes. Since the simplest solution is always the best, the notion that extraterrestrial forces directed Earth-bound processes 'violated [scientists'] parsimonious instincts'.[71]

Yet the idea kept cropping up. Towards the end of the seventeenth century, Edmund Halley had remarked to the Royal Society that the Bible's Great Flood was possibly caused by a comet. A century later, Pierre-Simon Laplace (1749–1827) whose modified version of Kant's nebular hypothesis remains current, thought that a large meteorite could kill off whole species.[72] Examining the fossil record, the naturalist Georges Cuvier (1769–1832) postulated 'a world previous to ours, destroyed by some kind of catastrophe'.[73] Over time, other scientists aired the possibility of mass extinction, and H. G. Wells (1866–1946), who studied biology under T. H. Huxley and advocated Darwinian theory, essayed it for a wider audience in his two-volume nonfiction *Outline of History* (1920), an international bestseller:

We do not know what jars and jolts the solar system may have suffered in the past . . . Some huge dark projectile from

outer space may have come hurtling through the planets
and deflected or even struck our world and turned the whole
course of evolution into a new direction . . . but this is a lapse
into pure speculation.[74]

In the 1970s Derek Ager effectively 'rehabilitated' catastrophes
by presenting them as relatively regular catalysts of the geological process, but to scientists from a variety of disciplines, mass
extinction was still not worthy of speculation.

That impacts and their consequences were so reluctantly
explored reflects the same adherence to accepted theories that
prevented meteorites themselves from being readily embraced as
significant phenomena. To assess the effects of major impacts it
was necessary to overcome attachments to familiar concepts but
also the territoriality separating members of different disciplines
and their suspicion of findings not originating in their own fields.
In the late 1960s, NASA produced the first colour photographs
of the whole, resplendently illuminated Earth, inspiring pride
and environmental awareness on the one hand while the 'pseu-
doscientific popular fringe' announced alien invasions and
apocalypse on the other. Scientists of every ilk may have wished
to distance themselves from sentimentality or sensationalism,
but the persistent denial of meteorites, first as extraterrestrial
objects and then as forces of cataclysmic change, betrays an
almost organic outrage that life on Earth should be subject to
such intrusions, as if it existed somehow on its own, suspended
not in space but in cautious considerations.

The theory that the 'age of the dinosaurs' ended as a result
of cataclysmic impact 65 million years ago passed from heresy
to orthodoxy over the next twenty years. Persuasive proof was
found in a thin geological stratum encircling the earth between
the layers corresponding to the Cretaceous (145–66 million
years ago) and subsequent Tertiary periods, the so-called 'K/T
Boundary'. The geologist Walter Alvarez gathered samples of
the peculiar stratum in Umbria and with his father Luis Alvarez,
a nuclear physicist, enlisted a colleague at Berkeley to analyse
it. Unheard-of quantities of iridium were found, an element

abundant in meteorites and space dust but rare on Earth. Samples gathered from the K/T Boundary at worldwide locations produced the same results. The Alvarezes' paper of 1980 which announced their belief that the dinosaur extinction was triggered by a massive asteroidal or cometary impact was hotly contested, not least because

End of the millennium stamp issued in Mali, 1999, depicting the Cretaceous mass-extinction event and the eruption of Mount Vesuvius of AD 79.

its authors had trespassed on palaeontological territory.[75] Luis Alvarez further insulted the naysayers by calling them 'stamp-collectors'. Habeas corpus was once more demanded: not the impact mass but its crater. Eugene Shoemaker proposed the crater in Manson, Iowa, which was adjudged to have been formed 9 million years too early. In 1990, *Houston Chronicle* journalist Carlos Byars helped point researchers towards Progreso, a Yucatán Peninsula port town. A Mexican oil company had recorded magnetic anomalies issuing from a colossal ring-shaped offshore depression there in 1950.[76]

In 1991, the crater named Chicxulub was identified as the *memento mori* of a world-wrecking impact. The object that punched an 85-km-wide and 30-km-deep hole in Earth's crust weighed over a trillion tons. Nowadays even schoolchildren will tell you that an asteroid was behind the dinosaurs' demise.[77] But the controversy surrounding the K/T extinction is not over, with Princeton University researchers presenting evidence that extensive volcanic activity (in India's Deccan Traps) had already done most of the work of killing off 70 per cent of Earth's species and that the Chicxulub impact only delivered the *coup de grâce*.[78] Either way, it is now accepted that at random points in time Earth experienced bombardments that altered planetary chemistries, presenting an evolutionary cul-de-sac for some types of life and favourable conditions for others, including us. Cosmochemists now speak of a 'biology of revolutionary rather than evolutionary change'.[79]

Investigations of the transformative power of large impacts indicate that even events much smaller than Chicxulub could produce enough dust and ash to block the sun and launch a global winter.[80] People living in desert cities know how a few gusts of wind can kick up a sandstorm so dense that visibility is reduced to near zero. Witnesses of volcanic eruptions have likewise watched ash clouds towering into contused skies, blotting out the sun. Those who have survived bombings can attest to the deafening shockwaves and lung-searing heat. But aside from Hiroshima and Nagasaki there is nothing in human experience that remotely resembles the sheer power, many orders of magnitude greater than any man-made weapon, of a cosmic impact, for which death itself is but a puny metaphor.

The speed of an object the size of the one that created the Chicxulub crater would compress the air beneath it, generating temperatures some ten times higher than the surface of the sun (60,000°C). An instant after hitting the atmosphere, the object would strike the ground and vaporize, shattering nearly 1 million cubic km of surrounding rock and creating a 1,000-cubic-km ball of white-hot vapour that would incinerate everything within a 250-km radius. Everyone within 1,500 km of the epicentre would be hurled to the ground amid a blinding flash of light. Earthquakes and tsunamis would follow as showers of burning debris falling back into the Earth's upper atmosphere set the world aflame.[81] Survival would be possible but not necessarily preferable. Scientists still debate whether the asteroid that created Chicxulub was 5 or 10 km wide, or whether a smaller, faster-moving comet was responsible, but given the scale of devastation the point seems almost moot.

Imagining destruction on such a scale remains essentially, perhaps mercifully, impossible but that hasn't stopped Hollywood from trying. Impact events have inspired a genre of popular film (and fiction) worthy of a book of its own.[82] That mass extinction should become a subject of mass entertainment is nonetheless suggestive, as are academia's jocularly titled contributions to the literature: works like 'Killer Rocks from Outer Space', 'Earth's Greatest Hits' and 'Coming Attractions'.[83] In u.s. military and scientific circles, the subject of planetary defence systems to intercept meteorite attacks is likewise subject to 'the giggle

factor', according to Air Force Lieutenant Colonel Martin E. B. France:

Even the most ardent supporters of defending the earth from cataclysmic asteroidal and cometary impacts share occasional public or private chuckles with colleagues and skeptics – behavior considered unthinkable when discussing means to avert or mitigate [terrestrial catastrophes].[84]

Victoria Crater, Mars, dia. 730 m, photographed by the High Resolution Imaging Science Experiment (HiRISE) camera on NASA's Mars Reconnaissance Orbiter.

If laughter, as has been said, is the confusion between 'yes' and 'no', then the 'GF', as Lt Col. France calls it, may be owed to the paradox of a threat so seemingly remote yet ever-present. 'Ours is indeed an age of extremity', wrote Susan Sontag in her 1965 essay 'The Imagination of Disaster', 'for we live under the continual threat of two equally fearful but seemingly opposed destinies: unremitting banality and inconceivable terror.'[85]

Over the last 200 years it is not the probability of impact-related catastrophe that has increased, but our awareness of it. Dismissing the danger of falling stones in 1803, a contributor to the *Journal des débats* confidently wrote that 'one should not worry; we are not at war with the moon . . .'.[86] A century later, the author of *Astronomy for Amateurs* assured his readers that 'there is little to fear of the destruction of humanity by these

Armageddon (1998), film poster.

balls of wind [comets]', based on the fact that two comets had been observed the previous century without incident.[87] In light of recent space surveys, we are now informed that Earth exists in 'a sort of cosmic shooting gallery' and the threat has been far from fully assessed.[88] Since NASA began monitoring Near-Earth Objects (NEOs) in 1984, the danger of mass extinction has been taken with greater gravity. The 'Spacewatch' programme was expanded in the 1990s and renamed 'Spaceguard' in response to a catastrophic event on Jupiter, the first of its kind ever predicted and observed from Earth.[89] Comet Shoemaker-Levy (named in honour of its discoverers, Eugene and Carolyn Shoemaker and David Levy) consisted of at least fifteen fragments, several of them over 1 km wide, which struck Jupiter with a cumulative force of 300 gigatons of TNT, underscoring the threat of a similar event to our planet.[90]

Earth has in fact experienced several near misses, one in 1991 when an asteroid was spotted only after it had passed at a distance of 170,000 km, 'in cosmic terms the equivalent of a bullet passing through one's sleeve without touching one's arm'.[91] On 15 February 2013 a NEO passed, this time as predicted, an order of magnitude closer to the planet's skin (17,200 km), darting gracefully through a flock of orbiting telecommunications satellites without grazing so much as an antenna.[92] But that same day an unobserved asteroid exploded above Chelyabinsk, Russia, causing considerable mayhem and prompting the U.S. House Science Committee to convene a meeting to discuss threats from space. In June 2013 NASA, which claims to have found 95 per cent of all large asteroids near Earth's orbit, launched its 'asteroid grand challenge', aimed at identifying the rest in cooperation with the private sector, academia and ordinary citizens.[93] As of 27 February 2015, 12,328 Near-Earth Objects have been discovered. Some 866 of these are asteroids with a diameter of approximately 1 km or larger. Also, 1,558 of the NEOs have been classified as PHAs (Potentially Hazardous Asteroids).[94]

Many scientists are now convinced that Earth has been subject to impacts '100 times more energetic [than Chicxulub] . . . with effects dwarfing that of the K/T', partially and perhaps fully vaporizing oceans.[95] Our earliest ancestors were life forms capable of taking refuge in the murkiest mulch at the bottom of the sea, places that coincidentally have yet to be as well-mapped as the surfaces of the moon and Mars. That *Homo sapiens* has overcome the cosmic odds to not only exist but examine its place in the universe must give pause, especially considering that the process was randomly detoured by flying rocks. The brilliant and eccentric cosmologist Sir Fred Hoyle (1915–2001) compared the chances of natural selection producing results like us to a whirlwind blowing through a junkyard and assembling a Boeing 747.[96] Whether or not evolution was aiming in our direction, its mind-bending trajectory demonstrates

Artist's interpretation of NASA's Wide-Field Infrared Survey Explorer (WISE) orbiting Earth to identify potentially threatening Near-Earth Objects as part of the NEOWISE project.

the relentlessly creative force that V. I. Vernadsky, author of *The Biosphere* (1926), called 'the pressure of life'.

Humanity seems somewhat less enthralled with creation than destruction, predicting apocalypses throughout history, producing them by the dozen through war and greed, and watching filmic enactments of them in our spare time. In February 2013, when one potentially city-smashing asteroid went relatively softly into the night and the impact of yet another caused injuries but no loss of life, there was no global elation, no music and handholding, no peace treaties or amnesties to celebrate not one but two miraculously narrow escapes. We cheer instead the Hollywood heroes that save Earth from sinister asteroids, even though such deliverance is frequent enough and quite real. The fantasy of science fiction, Sontag remarked, reflects genuine 'worldwide anxieties' but also 'serves to allay them'. Nor is the possibility of annihilation by impact the only source of disquiet, judging by the number of global warming-related 'eco-disaster' films lately on offer.[97]

Artist's rendition of a collision creating asteroid families, the parent bodies of most meteorites, based on observations made by NASA's NEOWISE project.

It may be argued that humanity vies with asteroids when it comes to wreaking planetary havoc. The French naturalist Georges-Louis Leclerc, Comte de Buffon (1707–1788) spoke of the 'realm of man' referring to humanity as a geological force.

Likewise, the Russian geologist A. P. Pavlov (1854–1929) referred to 'the anthropogenic era', a time when 'mankind became a single totality in the life of the Earth', capable of living wherever it pleased.[98] In 2008 a proposal was presented to the Stratigraphy

Commission of the Geological Society of London to make the 'Anthropocene' (the 'new human' epoch) a formal geological classification, in recognition of humanity's power to transform Earth to suit its needs. Whether dated to the development of agriculture or the invention of the steam engine, the Anthropocene is characterized by environmental devastation and species extinctions.

Vernadsky also saw humanity as a biospheric force, not in terms of the damage it had done but the good it might do. In his last work, 'Some Words about the Noösphere' (1943), he envisaged humanity's 'immense future . . . in the geological history of the biosphere'. Vernadsky conceived of Earth's biosphere as a living whole and attributed to science a similarly creative unity of purpose, capable of addressing 'the problem of the reconstruction of the biosphere in the interests of freely thinking humanity as a single totality'. He believed that an ever-growing body of knowledge would shape our planet for the common good, bringing it to another stage of development. 'The new state of the biosphere, which we approach without our noticing it,' Vernadsky wrote, 'is the noösphere'.[99] Described as 'a world of active and contemplative intelligence integrating scientific and artistic knowledge', Vernadsky's noösphere signalled a new role for humans, not as a force acting upon or above nature, but within it.[100] Emphasizing that human evolution is inseparable from that of the biosphere, he suggests that through the human creation of a noösphere, Earth acquires the means of becoming, as it were, conscious of itself. 'Man can and must rebuild the province of [its] life and thought, rebuild it radically in comparison with the past. Wider and wider creative possibilities open before [us] . . . Man is striving to emerge beyond the boundaries of his planet into cosmic space', wrote Vernadsky, who 'like all great scientists was a great optimist'.[101]

It may be said that the science of meteoritics, its history and expanding repertoire, belongs to the noösphere, serving as a portal to worlds of both material and ineffable significance, adding to the sum of our knowledge while hinting at all that remains beyond our ken. The geologist Ted Nield summarized the intrinsic value of this line of inquiry:

Whatever meaning we choose to derive from them, meteorites speak to us of the chance events within deep time in which we all have a common origin; and of the Universe, which, one way or another, is destined to reclaim our every atom.[102]

2 Fallen Gods

As an index of both contemporary preoccupations and primal instincts, it's hard to beat the response of eyewitnesses to a 'super-bolide' (large fireball) that exploded over the Russian Urals on the morning of 15 February 2013. Flashing as bright as a second sun, it streaked through the clear Siberian skies trailing a plume of pinkish smoke. The asteroid, whose original mass was around 10,000 metric tons, struck Earth's atmosphere at a speed of over 16 km per second, breaking into smaller pieces and releasing more than 30 times the energy of the atomic bomb dropped on Hiroshima.[1] The explosion created a deafening blast wave, shattering glass, damaging roofs and causing numerous injuries in the nearby city of Chelyabinsk.[2]

Within moments, people began communicating their experiences via Twitter and other social media sites. 'I am still young, I don't want to die!' one Elena Zimnukhova tweeted. 'Oh my god,' another witness wrote on a local web forum, 'I thought the war had begun.' Another Chelyabinsk resident was quick to establish her priorities: 'I first grabbed my cat and passport and ran outside.' A more sanguine citizen tweeted, 'Well, it is either a plane or a meteorite.'[3] Vladimir Zhirinovsky, leader of Russia's far-right Nationalist Party, reached other conclusions: 'It's not meteors falling. It's a new weapon being tested by the Americans', he claimed, although some people blamed the Chinese. The local newspaper *Znak* quoted a military source as saying the asteroid was fortuitously intercepted by a Russian defence missile that blew it to bits in mid-air, as evidenced by the trailing smoke. In the streets of Chelyabinsk people cried 'Judgement Day!' and a

John Martin, *The Great Day of His Wrath*, 1851–3, oil on canvas.

priest of the Holy Trinity Church said the exploding fireball mirrored 'how the Lord visited us on the day of the Christ Presentation in the Temple'.[4]

The event was recorded on dozens of dashboard cameras, a gadget installed in many Russian cars to provide visual proof in case of accident-related insurance claims. Footage from that crisp Friday morning features a long, empty stretch of asphalt flanked by snow-covered ground followed by the appearance of the fireball, in some cases so bright as to cause a total screen white-out. In one clip, the driver listens to a pizza commercial as the object crosses his field of vision before finally muttering, 'what the fuck'. People who were in town when the blast wave hit recorded the sounds of shattered glass, frenzied shouts and the cacophony of shock-triggered car alarms. Someone else compiled the dash-cam clips and set the images to soothing electronic music.[5] 'This all gives us reason to think', a deacon in the Church of the Transfiguration remarked. 'Is the purpose of our life just to raise a family and die, or is it to live eternally?'[6]

The Great Russian Meteor, as it is now sometimes called, afforded the opportunity to compare the effects of a significant meteoritic event on our fellows with those described in historical accounts. The atavistic fear felt in Russia (however mingled with the anxieties of citizens of a bellicose world) and the attempt to propagandize the event (to favour some and condemn others) were nothing new. The Russian fireball was not, however, believed to presage some disaster or announce the arrival of a great personage (at least so far), as others have been in the past. Priestly appeals to religious sentiment aside, the sense of an encounter with the mystical or divine that was once so commonly associated with meteors, comets, fireballs and meteorite falls seemed largely absent. But within months of the event, the Church of the Chelyabinsk Meteorite was established. Its founder, paranormalist Andrey Breyvichko, and his small group of followers, believe that the meteorite contains 'a set of moral and legal norms that will help people live at a new stage of spiritual knowledge development'.[7] This sort of response to a meteorite fall is mirrored throughout history.

Oriented Chelyabinsk meteorite, recovered by a local resident who sold it for nearly u.s.$100,000. This is the largest oriented meteorite originating from the spectacular fireball of 15 February 2013; 12 × 10 × 6 cm, 889 g.

Meteoritic phenomena in fact played a pervasive role in shaping the mythic narratives that elaborated the concept of the divine in concert with the observed world; they feature in the mythologies of ancient civilizations as well as the oral traditions of the native peoples of Australia, Africa and the Americas. Meteors and comets were seen as heavenly messengers interpreted variously as the harbingers of death or birth, victory or defeat, famine and drought or an abundant harvest. They acquired a political dimension as hagiographical embellishments, as the gods' warning or approval regarding weighty decisions or as arbiters of some great man's fate. The Spartan tradition of assessing a king's suitability at eight-year intervals involved keeping watch on a moonless night for a falling star, the sign of divine displeasure and the end of a royal career.[8] Virgil's *Aeneid* (29–19 BC) describes a scene in which Aeneas tries to flee a burning Troy (*c.* 1250 BC) with his son, Ascanius, and father, Anchises. His reluctant father looks to heaven for guidance and receives an omen in the form of what sounds, from his description, like an exploding fireball; 'a peal of rattling thunder' and

Mattia Preti, *Aeneas,
Anchises and Ascanius
Fleeing Troy*, 1635,
oil on canvas.

'a streaming lamp along the sky, which on the winged lightning
seemed to fly'. 'Now, now', cries Aeneas' father, 'no longer delay!'[9]

Livy's (*c.* 59 BC–AD 17) *History of Rome* lists portents and
prodigies, including meteors and meteorite showers that influenced both senatorial decisions and religious ritual.[10] Suetonius'
Lives of the Caesars (*c.* AD 119) speaks of a comet appearing several months after the assassination of Julius Caesar in 44 BC that
'shone for seven days . . . and was believed to be Caesar's soul'.
Tacitus' *Annals* (*c.* AD 116) mentions comet sightings under Nero
(*r.* AD 54–68) that Roman citizens, perhaps wistfully, interpreted
as presaging the emperor's death. Nero responded by slaughter-
ing a selection of nobles, ostensibly as a cosmic appeasement, but

also coincidentally eliminating some rivals. Emperor Vespasian's assessment of a comet passage during his rule (AD 69–79) was calmer: 'This bearded star concerns me not; rather should it threaten my neighbour, the king of the Parthians, since he is hairy and I am bald.'[11]

The Chinese may have been the first to record astronomical events beginning around 2000 BC with the observation of a solar eclipse. Mesopotamian astronomers likewise recorded 'celestial omens' from around 1200 BC, in the belief that 'history may repeat itself and that the gods [were] speaking to mankind in one way or another'.[12] Alongside the interest in singularities was the search for patterns, matching earthly affairs with heavenly ones in order to be armed for fate with knowledge. An understanding of the stars set men apart, as evidenced in the emergence of the prophet Zoroaster around 1100 BC. An early Christian text suggests that Zoroaster, 'a very great observer of the stars', used his wisdom to his advantage: 'wishing to be regarded as a divine being [he] began to elicit sparks from the stars and show them to people'. This brief passage and a story recorded in the first century AD have been interpreted as describing a meteor shower that Zoroaster may have anticipated.[13] The oldest portions of Avestan scripture, thought to record Zoroaster's words, say the sky is made of 'hardest stone' and worn as armour by Ahura Mazda, god of creation and cosmic order. Avestan texts contain many astronomical references, and the word *asana* means both 'sky' and 'stone'. On one occasion, Zoroaster was said to have defeated demons with 'a massive stone received from God'.[14]

Astonishment, as Martin Heidegger famously maintained, was

The Romans called comets *crinitas*, 'hairy stars', owing to their trailing tails. 'What Our Ancestors Saw in a Comet (after Ambroise Paré)', from Camille Flammarion, *Astronomy for Amateurs* (1903).

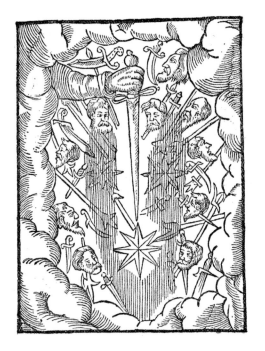

not only the beginning of philosophy but the force that 'carried and pervades'. The same might be said of myth. In *Language and Myth* Ernst Cassirer attempted to track the myth-making process back to its origin, beginning with 'momentary gods' arising from some anomalous experience:

> when external reality is not merely viewed and contemplated, but overcomes a man in sheer immediacy with emotions of fear or hope, terror or wish fulfillment: then the spark jumps somehow across . . . it is as though the isolated occurrence of an impression, its separation from the totality of ordinary, commonplace experience produced not only a tremendous intensification, but also the highest degree of condensation, and as though by virtue of this condensation the objective form of a god were created so that it veritably burst forth from the experience.[15]

Although conceived in a singular situation, the 'momentary god's' status evolves: Cassirer writes, '[He] achieves a certain substantiality which lifts him far above this accidental condition . . . he becomes an independent being, which henceforth lives on by a law of its own'.[16]

Common descriptions of exploding fireballs, with their blazing light and earth-shaking sounds, resemble the arrival of many a loud and fiery god, making them excellent candidates for the kind of process Cassirer describes. Meteoritic phenomena possessed the requisite tincture of astonishment to weave their way into myth, to be assigned attributes worthy of homage and to generate enduring beliefs and rituals.[17] Consider the arcane but irresistible impulse to wish on a falling star, a practice that may be traced to a time when the heavens were thought to be a dome the gods sometimes lifted to see what humans were up to. Should a star slip through, the story went, the wish of whoever saw it would be granted.[18]

There is considerable scholarly evidence that meteorites were venerated or worshipped in antiquity as the embodiment of gods. Sacred to the ancient Egyptians, the black, cone-shaped

'Ben-ben' stone kept in the temple of Ra at Heliopolis may have been a meteorite; the hieroglyphic signs used for 'bnbn' may be interpreted as indicating meteoritic iron and the stone's form is thought to have served as the prototype for later obelisk and pyramid capstones.[19] The Hittites of Asia Minor (with whom the Pharaohs waged two centuries of battle) used the 'black iron of the heaven' in rituals related to house-building.[20] In the Graeco-Roman period, temples were raised to at least nine deities described as 'fallen from heaven' and coins were struck in their honour, some bearing the image of a conical mass on a four-pillared shrine. The Greeks called these sacred stones *baetylos*, perhaps derived from the Hebrew *beth-el*, 'abode of the gods'. In Phoenician myth as recounted by Eusebius (*c.* AD 260–340), Baetylus is a son of the god Uranus who was said to have created the 'souled stones'.[21]

Meteorite worship was often associated with female deities. An incident during the thirteenth year of the Second Punic War (205 BC) involved the protective powers of a black stone revered in Pessinus (Anatolia) as a simulacrum of Cybele, the Anatolian 'mother of the mountain'. 'The actual statue of the goddess fell from Zeus, but no one knows what it is made of or who the craftsman was and they say it is not of human workmanship at all', wrote Herodian (*c.* AD 170–240) in his *Roman History*. The object figures in several versions of Rome's defeat of Carthage in 201 BC. It seems that Hannibal had crossed the Alps when a shower of stones deepened Rome's alarm. The senate called for a consultation of the Sibylline books (oracles believed to date to the sixth century BC) and found, according to Livy, 'that whenever a foreign foe should invade the land of Italy, he could be driven out and defeated if the Idaean Mother was brought from Pessinus to Rome'.[22]

An embassy was dispatched to King Attalus of Pessinus, whom Ovid portrayed as unwilling to part with the sacred stone until an earthquake shook his palace's foundation. A series of portents attended the object's arrival in Rome where it assumed the title of *Mater deum magna idaea*, 'great Idaean mother of the gods'. Debarking at Ostia in April 204 BC, the Magna Mater was passed from hand to hand until it reached the Temple of Victory on the Palatine Hill, 30 km and many odd hands away. An annual holiday was declared and incorporated into festivities surrounding the vernal equinox, including the Hilaria, when a pine tree was brought to the temple in honour of Attis, Cybele's young lover, amid 'wild, orgiastic rites' conducted by Phrygian priests who dressed and behaved like women.[23]

The Bronze Age cult of Astarte (Greek for the Mesopotamian-Semitic Ishtar), goddess of fertility, sexuality and war, may also have been associated with meteorites. Eusebius quotes a Phoenician source: 'travelling over the earth [Astarte] found a star that had fallen from heaven; she picked it up and consecrated it in the holy island of Tyre' (Lebanon), which became a cult centre. One of Astarte's symbols was a star within a circle signifying the planet Venus and the Greeks accepted her as Aphrodite. Tacitus (c. AD 56–117) described the worship of Aphrodite at the Temple of Paphos (Cyprus) where 'the representation of the goddess has no human form: it is a rounded block larger at the base and narrowing up to the summit, like a cone.'[24] A similar stone represented the sun god Elagabalus (from the Phoenician, and later Arabic, *el-gabal,* mountain) in Emesa (Homs, Syria) where, wrote Herodian:

Votive statue of Astarte, ceramic, *h.* 12 cm, from Pella in Jordan, dating to *c.* 800 BC.

> they built for him a very great temple adorned with gold, silver and splendid marbles. In this shrine there is no statue sculpted by the hand of man, and representing some deity, as with the Greeks and Romans; but one sees a very large stone, round at the base and ending in a point. Its form is

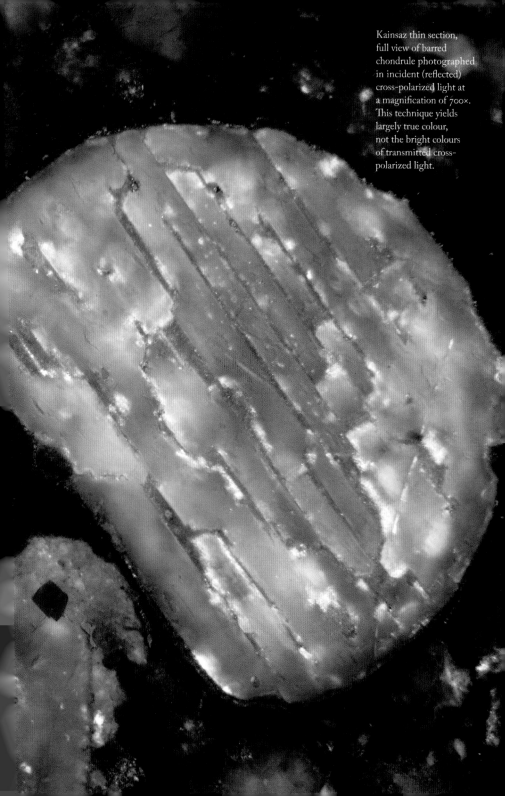

Kainsaz thin section, full view of barred chondrule photographed in incident (reflected) cross-polarized light at a magnification of 700×. This technique yields largely true colour, not the bright colours of transmitted cross-polarized light.

conical, its colour black. They [believe] this stone [fell] from Jupiter, [Zeus] and they show on it some small projections and cavities. They presume to see in it a rough representation of Helios [Sol].[25]

These references to cones and sun-like markings recall the morphology of 'oriented' meteorites whose shape is acquired during a relatively stable flight through Earth's atmosphere, some of which bear uniform marks of heat ablation or 'flow-lines' radiating back from the leading tip.

The Emesa stone profoundly influenced the life of Marcus Aurelius Antoninus Augustus (c. AD 203–222), who became Roman emperor at the age of fourteen thanks to his Syrian mother's manoeuvrings. A child priest in the Temple of Elagabalus at Emesa, he brought the sacred stone to Rome and installed it on the Palatine Hill near his palace, an event replicated during

Roman silver coin (AD 75–76) of Emperor Vespasian, showing the conical mass representing Aphrodite at Paphos (Cyprus).

The 2.7-kg Karakol chondrite LL6, which fell in Kazakhstan on 9 May 1840, shows 'sun-like' ablation patterns and is a fine example of an oriented meteorite (the 's' cube measures 2 cm).

A bust of the Roman emperor Elagabalus, *c*. AD 220.

his rule at the summer solstice festival. The young emperor began calling himself Elagabalus, and according to his contemporary Herodian, soon moved the stone to a sumptuous new temple:

[Placing] this god on a chariot adorned with gold and the most precious stones, he transported him from town to the suburb. [Elagabalus] thus led this car, which was drawn by six spotless white horses of the largest size, set off with much gold and variously caparisoned, and himself held the reins. Nobody ever mounted on the chariot ... [Elagabalus] ran backwards before [it], constantly looking at the god and holding the reins. And in order that he might not fall backwards, or slip, for he was not seeing where he was going, the way was strewn with gold dust.[26]

Aelius Lampridius, who authored a biography of Elagabalus, reports that the emperor sent for a statue of Astarte kept at Carthage, known as the Celestial or 'Astroarch' ('queen of the Moon'), to become his black stone's bride. 'For [Elegabalus] was saying that it became the Sun to espouse the Moon.' The teen emperor's behaviour, including an adventurous sexuality (dressed as a woman, he prostituted himself at Roman taverns) and sadistic pranks (such as drugging banquet guests and having them awaken in a room with ravenous beasts), made Caligula's extravagances pale in comparison. Elagabalus was eighteen when he and his mother were assassinated; the fate of the married meteorite remains unknown.[27]

Since none of the black stones mentioned in antiquity in connection with worship or veneration have survived, we cannot

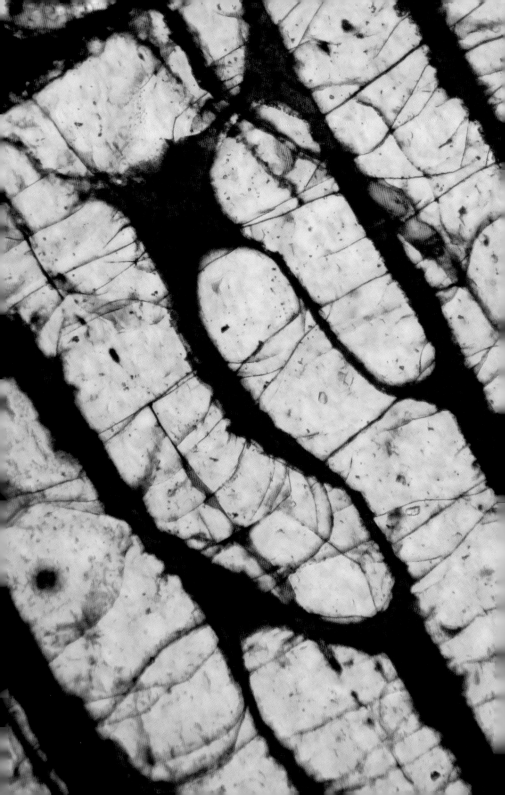

be certain they were meteorites; but aside from the matching descriptions there are more recent instances to be considered. A meteorite observed to fall in East Africa in 1853 was declared a god by the local Wanika people, who decked the 577-gram stone in pearls, anointed and clothed it. When a devastating attack by the rival Masai cast doubt on the meteorite's talismanic powers, the Wanika sold it to German missionaries.[28] In Andhra, India, two Brahmins saw a meteorite fall on 2 December 1880 and the site was soon engulfed with supplicants. According to A. Cunningham, who visited the area several weeks later, the stone was 'about the same size and shape as a common loaf of Indian bread . . . [and] named *Adbhuta-Nath*, "the astonishing or wonderful god"':

> A brick temple had already been begun . . . the walls were about two feet high. The votaries crowded in to make their offerings of flowers, sweetmeats, milk, rice, water, bel-leaves [bael], besides money both silver and copper. Two bel-trees close by had already been stripped of their leaves. After making their offerings the people knelt down in front and with joined hands muttered some prayers. One old woman, who seemed to be particularly earnest, even clasped the stone.[29]

Cast as middlemen at the liminal boundary of the material and metaphysical worlds, meteorites offered the desired proof of a higher power, one that could be seen, heard and touched.

Living close to nature, diverse North American native peoples venerated meteorites, and some still consider them sacred artefacts. The Camp Verde meteorite (a 61.5-kg iron from the Canyon Diablo strewn field near Meteor Crater, Arizona) was found wrapped in a blanket of feathers in a stone cist of the kind used to bury children (dated to around 1100–1200).[30] The Winona stony meteorite (24 kg) was found in a similar burial near Flagstaff, Arizona.[31] To the Pawnee of Nebraska, meteorites were the children of their principal deity, Tirawahat, and ritual fumigations of the stones were performed before battle.[32]

The barred chondrule in a thin section of the Dhofar 008 (Oman) meteorite, photographed in cross-polarized light at a magnification of 160×.

Edward S. Curtis, *The Mealing Trough*, Hopi women grinding flour, *c.* 1906.

The Hopi people of Arizona believed their gods came from the clouds, and that one of them made Meteor Crater his home. Some of their rites involved the white coesite found at the site, 'a rock flour almost as soft and fine to the fingers as talc' produced by the shock of impact on quartz-bearing rock.[33]

The Clackamas of northwestern Oregon could not have witnessed the fall of the Willamette meteorite, since it was carried to the place where they discovered it by gargantuan floods at the end of the last ice age, well before the area was inhabited. The largest iron meteorite found in North America (14.5 metric tons), it is riddled with cavities from water erosion and rust and known for the deep hollow ring it sounds when struck. The sound and the meteorite's otherworldly appearance may account for the Clackamas' name for the stone, Tomanowos, 'heavenly visitor' or 'visitor from the moon'. Clackamas youths were initiated beside the meteorite and prior to battle, warriors washed their faces and dipped their arrowheads in the rainwater gathered in its pitted surface.[34]

The Comanche, Apache and Kiowa peoples believed the Wichita County (Texas) meteorite came from the Great Spirit; well-beaten paths led to the 145-kg iron where beads, arrowheads and tobacco were left as votive offerings. In 1856, when Indian Agent for Texas Robert S. Neighbors sent a government wagon to take the meteorite away owing to its 'scientific importance', the Comanche begged that it remain. They gathered around it, 'manifesting their attachment by rubbing their arms, hands and chests over it'. Neighbors had the meteorite hauled to the state capitol in San Antonio, where it survived a fire that gutted the building in 1881.[35]

The Cree and Blackfoot peoples of Alberta, Saskatchewan and Montana believed that the Iron Creek meteorite, found on a hilltop near a tributary of the Battle River in Canada, fell from heaven. Considered a 'medicine stone', they called it Manitou, after their god; viewed at a certain angle, the 145-kg meteorite resembles a man's profile, and food and trinkets were offered to it prior to hunting expeditions and battles. Perhaps owing to concerns that such superstitions would impede the spread of Christianity, the meteorite was transported to a nearby mission house in 1870. A medicine man's prediction that 'evil will sweep over the land . . . sickness, starvation and devastation' if the god left his people's possession came true during an epidemic of smallpox, a disease brought by European settlers, that winter.[36] 'Ever since that rock was taken, [catastrophe] is basically what

Edward S. Curtis, *Storytelling*, group of Apache men, *c.* 1906.

happened to our people', said the president of the Blue Quills First Nations College, St Paul, Alberta, in 2012, requesting that the Royal Alberta Museum (Edmonton), which acquired the stone in the later 1800s, relinquish the meteorite so that it might again feature in traditional ceremonies.[37]

A similar belief was expressed in southern Nigeria, where the Uwet meteorite that fell in 1825 was considered beneficial to the local community. The British authorities who removed it in 1903 were blamed for an outbreak of smallpox that occurred shortly afterwards. The authorities were obliged to return it and the community's recovery was attributed to the meteorite's talismanic presence.[38] More recently, the Moqoit people of Argentina foiled attempts to ship the 37-ton El Chaco meteorite to Germany to feature in the 2012 Documenta art show. El Chaco is the largest single meteorite from Campo de Cielo ('field of the sky'), the world's vastest strewn field that has yielded thousands of tons of meteorites and comprises numerous craters. The impacts that caused them probably happened around 2000 BC, and figure prominently in the Moqoit's cosmology. Argentine scientists rallied to the Moqoit cause, stating the meteorite was not a 'cosmic curiosity' but a cultural and natural artefact best viewed where it landed, in Chaco Province.[39]

Once dismissed as whimsical, geomythology, the study of how geological events or features were incorporated into traditional beliefs, has gained credence owing partly to research related to meteorites and impact craters. Viewing myths as more than a manifestation of creative imagination but as possibly containing verifiable information, astrophysicist Duane W. Hamacher discovered an impact crater in Palm Valley, Australia (130 km from Alice Springs), in 2009. His search was inspired by a tale told by the Arrernte Aboriginals of a star that fell to earth in a waterhole, 'where the serpent Kulaia lived, making a great noise like thunder'.[40] Comprising hundreds of groups, each with a distinct language, Australian Aboriginals have a history reaching back tens of thousands of years and oral tradition features prominently in their cultures. The crater Hamacher identified

Stan Gaz, *Wolfe Creek Crater, Western Australia,* 2009, solarized black-and-white print. Impact crater identified by geologist Frank Reeves in 1947, dia. 875 m.

predates the human presence on the continent, yet there are indications that other impacts may have been witnessed and the stories handed down across generations.

An Aboriginal name for the Henbury Craters (formed in Australia's Central Desert between 4,200 and 1,900 years ago) roughly translates as 'sun walk fire devil rock'. People preferred not to camp near the craters and refrained from drinking water gathered there, lest 'the fire devil fill them with iron'.[41] A similar fear is expressed by Aboriginal artist Jane Gordon with respect to the origin of the Wolfe Creek Crater, formed some 300,000

years ago: 'The people hid in a cave because the ground shook real strong. They saw dust coming up from the ground . . . They never touched that area. The star that fell down was evil.'[42] Stan Brumby, another artist, relates this version:

Jane Gordon, *Ngurriny* (Aboriginal name for Wolfe Crater environs), 2002, acrylic on canvas.

> Well, this star, he got slack. He got slack from a big storm that stretched from ocean to ocean . . . that star, him falling, falling, till, bang, he hit the ground . . . and that star pushed through to the underground water . . . big underground river . . . No one told me that story . . . I dream all about it in my sleep.[43]

Daisy Kungah,
The Star and the Serpent,
2002, acrylic on canvas.

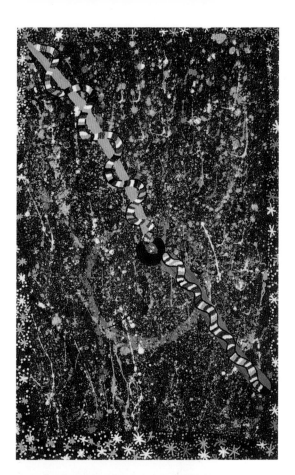

Stan Brumby, *The Sugar
Leaf Dreaming*, 2000,
acrylic on canvas.

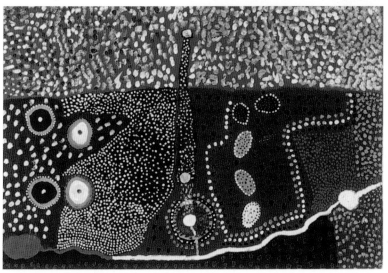

Artistic renderings of the 875-m-diameter Wolfe Creek crater feature the 'rainbow serpent . . . that snake that travels like stars travel in the sky', and is associated with a waterhole that formed in the crater's centre.[44]

Gosse's Bluff, 200 km west of Alice Springs, the 150-metre-high rim of a 145-million-year-old impact crater, is sacred to the Arrernte people. According to their oral histories, a 'sky-woman' dancing in the Milky Way accidentally dropped the wooden basket holding her child and its crash to Earth formed the high circular bluff.[45] People living near Tenham Station, Queensland, where a meteorite fall was recorded in 1879, were 'deadly afraid' of the stones, and covered them with 'kangaroo grass' and tree boughs to hide them from the sun, hoping this might prevent further falls.[46] Some Aboriginal groups see comets and/or meteors as malevolent spirits called 'eye-things', which scan the earth for souls to devour.[47]

The indigenous San people of southern Africa also consider meteorites dangerous: 'They can kill people, and at the times of the meteor showers when many are moving about and falling, the sky is very bad.' Echoing themes from Mesopotamian and classical antiquity, the San god Koaxa, 'lord of the animals', used a meteorite to fight lions that attacked his son. Like some Australian Aboriginals, the San associate 'menacing water snakes' with rivers and waterholes, only theirs emit light from their foreheads and are thought to represent fireballs or comets.[48] The serpents and snakes in San and Australian aboriginal portrayals of meteoritic events recall medieval European lore, where dragons may represent the same phenomena, with their burning eyes, field-scorching tails and thunderous roars.[49]

Some traditions related to meteorites seem designed to convey a warning. Among the Bororo people of southern Brazil, meteorite falls and exploding fireballs are believed to be capable of stealing human possessions or souls. Such events elicit a ceremony overseen by the village shaman, and have been described by one observer as 'having the quality of a cathartic outbreak':

San or 'Bushman' rock painting depicting a bolide, serpent, a thread of light and therianthropic figures, found in the foothills of the Maloti Mountains, South Africa.

Everyone, women and children down to young toddlers, sets off a chaotic uproar. People scream urgently, fire off guns, beat on pots, pans, and logs, crash rolled-up mats against the ground . . . the din was impressive, and nearly everyone was acutely anxious, on the verge of panic.[50]

The ritual seems to pay homage to a fearsome force as an appeasement to ward off the dangers it represents. This response to frightening but not fatal events may indicate that at some point in time, the Bororo or a related people survived an injurious meteorite fall, an experience they wanted no one to forget.[51]

Catastrophes often pass into oral or written history as cautionary tales, but their scale and consequence compel some individuals to seek the truth that may lurk behind the legends. Physical proof of the Bible's flood, God's sweeping punishment of mankind's waywardness, has been sought since the dawn of archaeology. Even now, archaeologists are searching for the remains of Solomon's Temple, described in the Old Testament as a magnificent edifice razed along with most of Jerusalem by

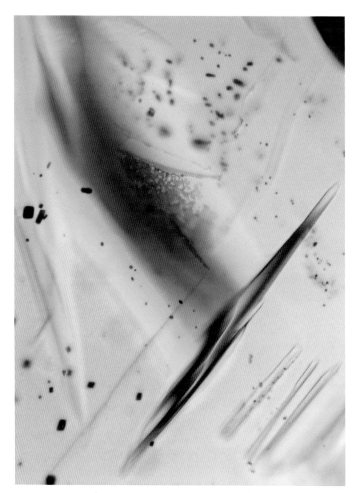

Thin section viewed at 160× magnification in cross-polarized light with the addition of a 1/4 wave retardation filter. Dr Tony Irving, who classified this unusual meteorite (found in 2010) as an 'ungrouped achondrite', noted its 'cumulate igneous texture and trains of tiny fluid-like bubbles'.

Nebuchadnezzar II, unwitting agent of a neglected and angry Hebrew god. The Qur'an tells a similar tale about the city of Ubar, levelled by God's wrath with a blast of wind when its people (the Beni 'Ad) abandoned him for their old idols. Warned by the prophet Hud to 'worship none but God' or face the consequences, the sceptical Beni 'Ad replied:

> Bring on us now the woes you threaten if you speak the truth. 'That knowledge [said Hud] is with God alone. I only proclaim to you the message with which I was sent. But

I perceive you are a people sunk in ignorance.' So when they
saw a cloud coming straight for their valleys, they said: 'It is
a passing cloud that shall give us rain'. 'Nay' [said Hud], 'it
is that whose speedy coming you challenged – a destructive
wind wherein is an afflictive punishment – it will destroy
everything at the bidding of its Lord!' And at morn, naught
was to be seen but their empty dwellings.[52]

In 1918 a Bedouin guide told British explorer Harry St John
Philby (1885–1960) that Wabar (Philby's transliteration of
Ubar) was deep in the forbidding Empty Quarter, which at
nearly 650,000 sq. km is the largest contiguous sand sea on
Earth. Bedouin lore spoke of an iron mass 'as big as a camel's
hump' near the place some called *al-Hadida* (Arabic for 'iron')
and of the 'blackened pearls of the [Wabar] king's ladies' also
found at the site. Convinced the ruined city existed along with
potentially priceless artefacts, Philby spent the next fourteen
years trying to obtain Saudi king Ibn Saud's assistance to organ-
ize a caravan to explore the Empty Quarter, an arduous journey
of several months over a merciless terrain.

In January 1932, when Philby left Riyadh with fourteen
Bedouin and 32 camels carrying provisions for 75 days, his first
objective was Wabar. Reaching the spot took a tortuous month;
without the 'incredible, altogether inexplicable' navigation skills
of Philby's guides, the entourage would have perished. The briny
wells that sustained the caravan could be a ten-day march apart
and water was so precious the Bedouin poured it sparingly into
the camel's noses (a technique called 'snuffing') in order to 'cool
their brains'. Having crossed countless undulating dunes, Philby
reached the top of a blackened crest and looked down:

> not upon the ruins of an ancient city, but into the mouth of
> a volcano, whose twin craters half-filled with drifted sand lay
> side by side surrounded by slag and lava outpoured from the
> bowels of the earth . . . .this may indeed be Wabar, of which
> the Bedouin speak, but it is the work of god, not man.[53]

Philby saw how the site's charred rim might have sparked the legend of the lost city, '[its] black walls stood up gauntly above the encroaching sand like battlement and bastions of some great castles.' He happened to have an issue of the Royal Geographical Society journal that contained an article mentioning the possible impact origin of a crater in West Africa. Philby duly noted in his diary that the craters he had found were 'perhaps depressions created by the fall of a meteorite'. He mapped the site (the largest crater was 91 m across and 12 m deep) and found a disappointing 'rabbit sized' iron. The men meanwhile collected loads of 'little jet-black shining pellets', the strange glassy globules they thought were pearls. Back in London, analysis proved the 'rabbit' was a meteorite.[54]

Nearly 30 years later, Thomas J. Abercrombie, a *National Geographic* magazine author and photographer in Saudi Arabia researching a story, met James Mandaville of Aramco Oil, a Saudi-American venture that had been exploring the Empty Quarter for decades. Mandaville told Abercrombie that he had recently heard that the capricious sands of the Empty Quarter had uncovered the camel-hump iron. Abercrombie organized an expedition in 1965, found the mass and published the scoop. 'The great nugget rang like a bell as I chipped a sample off the edge', he wrote. The 2.43-ton meteorite was later transported to King Saud University in Riyadh and placed on display.[55]

In 1998 impact-crater experts Eugene Shoemaker and J. C. Wynn organized expeditions to investigate Wabar, a trip by Hummer convoy from Riyadh that took just seventeen hours. They identified three craters, now all nearly filled to the brim with sand. Philby had not discovered a lost city, but he was the first to map 'the best-preserved and geologically-simplest meteorite site in the world'.[56] Rather than rock, the Wabar meteorite struck sand and sandstone, producing 'sheets of incandescent fire … a curtain of hot fluid, a mixture of the incoming projectile and local sand' that coated white ejected sandstone in reddish-black glaze, entirely transforming other pieces 'into a frothy glassy material so light it could float on water'. Wynn and Shoemaker described the formation of the impactite rock and glass pellets

(Philby's 'slag and lava' and the Bedouins' 'pearls') in detail and tentatively dated the craters as less than 450 years old.[57]

Supposing they were right, a plausible theory regarding the origin of the famous Black Stone, venerated by Muslims since the seventh-century founding of Islam, is disproved. Long rumoured to be a meteorite, the Black Stone of Mecca has never been analysed and its origins remain the subject of speculation. A paper of 1980 argued that it was impactite glass from the Wabar site, based partly on descriptions and reports of its ability to float.[58] Wynn and Shoemaker's dating of the fall would preclude this, since the Black Stone's history is centuries older.

Muhammad ibn Ishaq (*d.* AD 767 or 761), one of the Prophet Muhammad's biographers, tells how a fire partially destroyed the pre-Islamic Kaaba, a sanctuary in Mecca purportedly containing some 360 idols, including the Black Stone, in AD 605. The local clans could not decide who deserved the honour of repositioning the sacred stone in the refurbished sanctuary and agreed that the next man to walk by would decide. That man was the future Prophet Muhammad, who was 35 years old at the time and as yet untroubled by divine revelations. Muhammad said that the stone should be placed in a cloth and, with the four clan leaders each holding a corner, carried to the Kaaba, where he himself would set it in the building's eastern corner. The Kaaba became the

Pilgrims kiss the Black Stone in Kaaba, Mecca, Saudi Arabia, postcard, *c.* 1970s.

Hand sample of NWA 5192, found in Morocco in 2006, polished to 1/4 micron and photographed in incident light at a magnification of 345×.

centrepiece of Islam and the pilgrimage to Mecca became a religious obligation for all Muslims that involves touching or kissing the stone in homage to the Prophet. Tradition holds that it was originally white but was turned black by sinners' kisses.

Edward Sale, who spent 25 years in Arabia and in 1734 became the first person to translate the Qur'an into English, recorded beliefs related to the Black Stone:

Thin section (30 microns) of the Kainsaz meteorite, carbonaceous chondrite, from an observed fall in Tartarstan, 1937. Image taken in cross-polarized transmitted (pass-through) light at a magnification of 760×.

[It] is considered by some as the right hand of God on Earth; and they fable that it is one of the precious stones of paradise; which fell down to the earth with Adam. The most marvellous

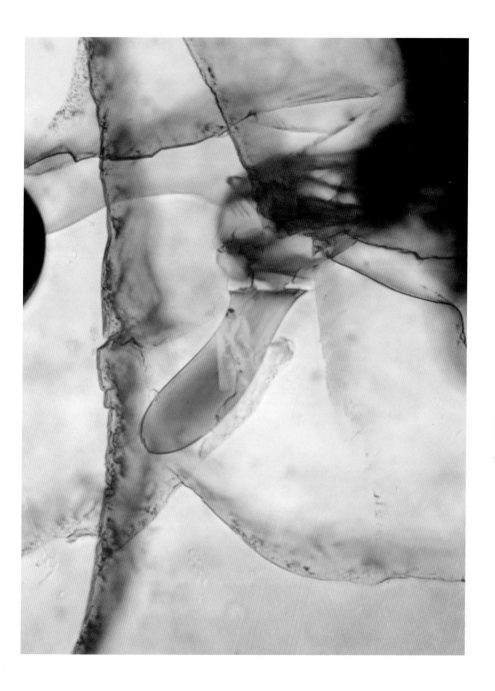

properties are attributed to this *aerolith*: that it floats in water;
that it instantly fattens the leanest camel which bears it;
and that such is its weight that several horses cannot move it
at all.[59]

Paul Partsch, curator of the Vienna Museum of Natural
History meteorite collection, published the first history of the
Black Stone in 1857, drawing on nineteenth-century travel
accounts. He also corresponded with Anton Ritter Von Laurin,
Austrian General Consul in Cairo, who was a friend of Egyptian viceroy Muhammad Ali. Ali showed Von Laurin what he
claimed was the only fragment of the stone outside of Mecca,
wrested from the fanatical Wahabis who had attacked the
Kaaba believing its contents to be idolatrous. Von Laurin noted
the stone's black surface and finely grained silver-grey interior
embedded with bottle-green cubes, a description consistent
with some meteorites.[60]

Like other storied artefacts the Black Stone was an occasional
source of contention between those who wished to destroy or pro-
tect it for religious or political reasons; it was repeatedly broken,
mortared together and bound with a silver hoop. Perhaps the greatest indignity it suffered was its unfortunate appearance on a Saudi
Arabian postage stamp.[61] The mystique of Islam's Black Stone
remains nonetheless intact. Unless its custodians one day permit
its scientific analysis, people are unlikely to stop believing (and
attempting to prove) it is a meteorite.

Tales of meteorite veneration maintain a peculiar grip on the
popular imagination, as evidenced during the autumn of 2012,
when the so-called 'space buddha' made worldwide headlines.
The 23-cm-tall figure of a man carved from meteoritic iron was
allegedly found during a Nazi expedition to Tibet and brought
to Germany in the late 1930s. Researchers at Stuttgart University tentatively described it as an eleventh-century Bon culture
(Tibet) representation of the powerful personage Vaiśravana,
'king of the north'. The seated figure bore a swastika (an ancient
symbol co-opted by the Nazis) on its chest and a halo around its
head and showed traces of what appeared to be gilding. Noting

80

that the Tibetans called meteoritic iron *namchang* ('sky iron'), researchers remarked that if the meteorite was indeed a thousand-year-old cult object it would be 'absolutely priceless'.[62]

Analysis proved that the figure was hewn from a fragment of the Chinga iron meteorite that fell on the frontier of Mongolia and Siberia some 10,000–20,000 years ago. But specialists in Tibetan art soon debunked the Bon/Buddhist origin, since holy men of these cultures are never depicted wearing trousers, shoes, cowls, beards or earrings in one ear like the 'iron man', as the figure is now known. Nor was there proof that the Nazis had anything to do with it, according to German historians who failed to find the figure listed on an inventory of objects recovered during the ss expedition.[63] Although the carving has been dated to some time in the twentieth century, the questions of where and why it was made and how it came to Europe have yet to be answered. The excitement surrounding the iron man is, however, no mystery. Forged from a meteorite, it was a 'religious statue with a stronger connection than most to the heavens'.[64]

Cabin Creek

Johnson Co. Arkansas.
27. März 1886.

3 To Have and To Hold

A meteorite's power to create a commotion was proved yet again on 7 August 1996 when u.s. President Bill Clinton stepped onto the South Lawn of the White House to comment to the media on a scientific discovery of dizzying portent. After years of research, NASA had published its findings regarding what appeared to be evidence of fossilized microbial life embedded in a meteorite from Mars named ALH84001, recovered during a u.s.-funded mission to Antarctica.[1] While cautioning his global audience that the research team's results 'would and should continue to be reviewed, examined and scrutinized', Clinton was obliged to note the prospects presented by the appearance under high-powered electron microscope of something closely resembling a minute version (100 times smaller) of a rod-shaped bacteria known to have existed on Earth 3.5 billion years ago:

> Today, rock 84001 speaks to us across all those billions of years and millions of miles. It speaks of the possibility of life. If this discovery is confirmed, it will surely be one of the most stunning insights into our universe that science has ever uncovered. Its implications are as far-reaching and awe-inspiring as can be imagined . . . We will continue to listen closely to what [rock 84001] has to say.[2]

Cabin Creek meteorite, notable for its regmaglypts, surface ablations acquired as it passed through Earth's atmosphere. From an observed fall in northwest Arkansas on 27 March 1886.

Never has a meteorite started so many conversations or posed so many questions. If Earth was not the only planet to have hosted life, how could we be sure it originated here? If it was instead delivered from Mars, would that make us Martians? Supposing Earth did not hold the patent on life, wasn't it possible that other kinds of biogeochemical geneses had taken place elsewhere, under different circumstances? Had we not sold the universe short by claiming it all for ourselves, and was it about to open to us in all its staggering multiplicity?

Clinton transmitted a more measured message from the meteorite, calling the results of its study 'another vindication of America's space program'. Whatever the outcome of the research regarding its evidence of extraterrestrial life, ALH84001 had effectively endorsed NASA's efforts and given the Pathfinder mission to Mars planned for later that year a terrific public relations and funding boost. The president barely hinted at the larger issues. Finding conclusive proof of life on other planets would spark a fire sale on preconceptions, the end of all anthropocentric preoccupations, such as religion and philosophy, and the beginning, no doubt, of new ones. The puerile politics of nationalism, 'the measles of mankind' as Einstein called it, would be unmasked and a more responsible globalism would be the order of the day. NASA's provocative findings did not, however, prove conclusive and humanity was spared this great effort.

Instead, several months after Clinton's announcement, in a high-ceilinged room on Park Avenue, spotlights played over glass cases provided by Cartier, where three Martian meteorites sat atop plush black velvet mounds like many a fabled jewel before them. Guernsey, a Manhattan auction house which estimated the three stones' value at U.S.$1.5–2 million, was not the first to place meteorites on the block. Phillips, another New York house, had done so the previous year as part of a more sober natural history offering (including fossilized dinosaur eggs) that garnered significant media attention. Thanks to ALH84001, the Guernsey auction of November 1996 attracted far more.

'People were again faced with the question', wrote Arthur Hirsch for the *Baltimore Sun*, 'that has haunted mankind since the dawn of consciousness: What's in it for me?' . . . [the Guernsey auction] reminds us that in America no frontier of human experience or imagination is so cosmic it cannot be packaged, hyped and sold.'[3] The Martian samples failed to meet bidding expectations but the auction-related hype planted a kernel of fascination in some fecund minds. The release in 1998 of *Deep Impact* and *Armageddon*, Hollywood blockbusters about the heroic and ingenious destruction of threatening near-Earth objects, heightened meteorites' post-industrial sheen. In parallel with the NASDAQ market surge, tech-savvy young investors began to 'diversify into meteorites', making them the 'collectible of the moment' and driving prices sky high before the bubble burst.[4]

The ALH84001 sensation was not entirely responsible for the surging interest in meteorites, which had been growing since the 1980s controversy over the Cretaceous mass extinction, but it threw a recognizable pattern into relief. The rise of meteorites as a commodity in the 1990s in many ways mirrors the nineteenth-century frenzy of acquisition aroused by the discovery of ancient Egyptian artefacts. Just as twentieth-century space-age technology revealed the secrets entombed in meteorites, so the mid-1800s invention of photography and the nascent discipline of archaeology had alerted the public to the treasures of bygone civilizations. Museums of every great capital, scrambling to enrich their antiquities' collections as advertisements of their cultural ascendancy, likewise sought to expand their cabinet of meteorites to demonstrate their mastery of science.

Collecting antiquities was an imperialist contest, especially between the French and British who fought for possession of Egypt's Rosetta Stone and bickered over who had the bigger obelisk. The consular agents dispatched to acquire artefacts for museums and private collections were adventurers, not diplomats; their work demanded shrewdness, resourcefulness and a rugged constitution. Transporting king-sized antiquities involved backbreaking manoeuvres and feats of engineering considered well worth the effort, since monuments such as the obelisks raised beside the Thames in London, on Paris's Place de la Concorde and in New York's Central Park enhanced both urban aesthetics and national prestige.[5]

Collecting meteorites sometimes involved similar exertions alongside a degree of patriotic competition. When the founding

director of Vienna's Imperial Geological survey, Wilhelm Haidinger, sent a meteorite that fell in Romania in 1858 to the Vienna Museum of Natural History, his accompanying note hailed the collection, the world's first, as 'a veritable jewel, a manifestation of the zeal, knowledge and perseverance of our Vienna, our Austrian fatherland!'[6] Cash-strapped museum curators vied for prized meteorite specimens and kept a close count of who possessed what. In 1896 Vienna and London were neck-and-neck, with the former boasting 482 and the latter 476 specimens. America lagged behind. The Smithsonian did not begin collecting meteorites until the 1880s and the American Museum of Natural History had just nineteen specimens until J. P. Morgan purchased and donated an extensive private collection in 1900.[7] No self-respecting meteorite exhibit was complete without a ponderously large exemplar. The British received a 634-kg Campo del Cielo iron as a token of appreciation from the Argentine government for having acknowledged their independence

The meteorite collection at the Natural History Museum, Vienna.

from Spain. A frigate brought the prize to London in 1826, where it became the first large meteorite to be exhibited at the British Museum.

The American Museum of Natural History staked a claim on the monumental Cape York (Northern Greenland) meteorites, whose existence was first alluded to in a report of 1818 by a British officer, Captain John Ross. Having noted that the Inuit people of the area used harpoons with strange iron tips to kill whales, Ross collected samples that turned out to be made of meteorite, laboriously hacked from a larger mass and hammered into shape. But Ross was unable to find the source of the meteoritic iron, as were several subsequent expeditions. It was not until the 1890s that American navy captain Robert E. Peary located three large meteorites on the northern shores of Melville Bay during his Arctic explorations. He had failed to enlist the Inuit's assistance during previous searches but Peary, who had crossed Greenland in a dogsled, won their respect and was led to the site by a local guide. Peary was intent on reaching the North Pole, a race involving the representatives of several nations, not unlike the rush to the moon less than a century later. Yet over the course of several years he found time to load and ship the meteorites the Inuit named 'Woman', 'Tent' and 'Dog' to America, an excruciating labour in the frigid fastness that involved rafting the meteorites on an ice floe to the waiting ship.[8]

Robert Peary aboard the ss *Roosevelt*, c. 1909.

Moving Dog (400 kg) was easy. Woman (3 metric tons) was hoisted on a timber sledge with iron rollers on a wooden track and offloaded onto a raft of thick ice that gave way an instant after the mass was winched aboard the ship. The Tent's (31 metric tons) transport was the most demanding.

During one expedition its sunken bulk was excavated; during the next it was pried from its ancestral pit, jacked onto the improvised railway that carried it to the lip of the coast and rolled down the slope to the bay, where it rested until the following year. Peary returned in 1897 and with the Inuit's help built a wooden bridge fitted with iron rails from the shore to a carefully counter-weighted ship. That autumn, Tent arrived safely in the Brooklyn Navy Yard, where it rested for seven years until Peary sold all three Cape York meteorites for U.S.$40,000 to the American Museum of Natural History in uptown Manhattan, where they remain on display today.[9]

Just as the growing demand for ancient artefacts had prompted a reassessment of their value, resulting in legal decrees as early as 1835 (in Egypt) to prevent plunder, so meteorites became the subject of legislation as their desirability to museums and collectors became apparent. A precedent-setting case involved the Willamette meteorite in Oregon; abandoned by the Clackamas when they were forced from their land, it was found by a farmer named Ellis Hughes in the woods a little over a kilometre from his house. He and his teenage son spent three months transporting the 15.5-ton mass home through the dense

In 1902, the Willamette meteorite was ingeniously transported from the property of Oregon Iron & Steel Co. to that of the meteorite's finder, and neighbouring landowner, Ellis Hughes.

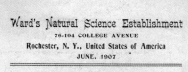

Ward's Natural Science Establishment
76-104 COLLEGE AVENUE
Rochester, N. Y., United States of America
JUNE, 1907

Meteorites For Sale

Willamette Siderite, lower side

Since the last edition of this price list was published there
have been very extensive changes in our stock of meteorites.
Special attention is called to the new falls represented, including Coon Butte, Elm Creek, Estacado, Modoc, and Shel-
burne, and to those which hitherto have not been on the
general market, such as Billings, Santa Rosa, and Willamette.
Practically all of the specimens of the rarer meteorites that
we have in stock are mentioned in this list, but in the case of
the more common ones, such as Cañon Diablo, Toluca, Ness
County, Pultusk, etc., quotations are given on but a small
portion of the specimens we have, and prices on other weights
than those listed will be furnished on application.

Collections of typical but inexpensive meteorites are
described in a special circular which will be sent on application.

Pioneering meteorite collector Henry Ward (1834–1906), who founded one of the earliest commercial suppliers of fossils, minerals and meteorites for museum, university and private collections, authenticated the extra-terrestrial origin of the Willamette. Ward's Natural Science Establishment catalogue, 1907.

brush. They built a robust timber cart with tree trunk-sized wheels, jacked the 3-metre-long mass atop it and rigged wire cables from the cart to a makeshift capstan anchored to trees and turned by horses. Having hauled it home, tree by tree, Hughes and son built a shed around the meteorite and did a brisk business charging people 25 cents admission.[10]

Word of the attraction reached the Oregon Iron & Steel Company, who said Ellis had found the meteorite on their land and therefore claimed ownership of it. When Ellis refused to relinquish the hard-won meteorite, they took him to court. Public opinion favoured Hughes for his industrious-ness but the company won. Oregon Steel promptly sold the Willamette to the mysterious Mrs William E. Dodge 11 of New York (for U.S.$26,000), who in turn donated it to the American Museum of Natural History in 1905 with the proviso it remain intact.[11] Nothing else is recorded of this generous patron, whose wishes were ignored in 1997 when the meteorite's crown was cut from the mass, enabling museum experts to examine the Willamette's internal structure and to trade the severed portion for other specimens, including a coveted piece of Martian meteorite.

Laws governing antiquities are still applied to meteorites. In the United States, large ones are covered under the Antiquities Law of 1906: if found on state or federal land they become government property, if on private land, they belong to the land-owner. Small meteorites were once freely prospected on federal land, like fossils and minerals, but owing to the combined de-mands of scientists, hobbyists and commercial collectors, official permits are now required. Some countries forbid the export of

Willamette meteorite in the American Museum of Natural History. The circle indicates where a small portion was severed.

meteorites. In 1970, 90 nations signed a UNESCO resolution that allows for their recovery if illegally trafficked, the same as is agreed for other 'cultural properties' including antiquities.[12] Meteorites have indeed been called 'the ultimate antique', at 4.5 billion years old and counting, they are the oldest thing one can own.[13] Just as artefacts of ancient civilizations help reconstruct humanity's past, so meteorites shed light on the origin of our planet and solar system. And just as antiquities collectors, at first content with any hoary relic, soon sought to acquire the finest examples of ancient artistry, so have meteorite collectors refined their tastes. In addition to being something from long ago and far away, select meteorites are now considered *objets d'art*, and command prices in the hundreds of thousands of dollars.

'Beyond matters of the soul, the inspiration for most art is in nature', says collector Darryl Pitt, who placed several choice meteorites on the Phillips auction block in 1995, calling them 'natural art from outer space'. The Detroit-born Pitt developed a fascination with meteorites at the age of thirteen when he visited

Meteor Crater in Arizona. Influenced by the art theorist Rudolf Arnheim, one of his college instructors, Pitt was captivated by the sculptural forms of some meteorites, likening them to the works of Giacometti, Henry Moore and Barbara Hepworth.[14] At the vanguard of meteorite collecting in the late 1980s, Pitt is now the primary owner and curator of the Macovich Collection, boasting 'the finest aesthetic meteorites in the world', including rare specimens of lunar and Martian meteorites acquired through a variety of avenues, such as trades with major museums, collectors and dealers worldwide. In distinguishing aesthetic from 'prosaic' masses, Pitt created an art market for meteorites and effectively branded them, with those bearing the Macovich label attracting a roster of stellar buyers, including film-maker Steven Spielberg, Bruce Willis (hero of the film *Armageddon*) and Qatari prince Saud bin Mohammed al-Thani.

Left, Gibeon iron (Namibia) and right, Mundrabilla (Western Australia), sculptural meteorites referred to as 'tabletop specimens'.

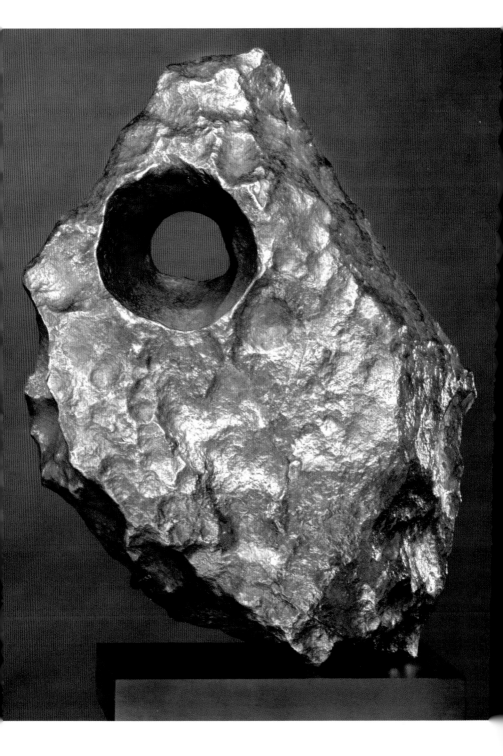

This Gibeon iron (27 × 19 × 8 cm, 12.38 kg) resembles *Single Form*, 1964, a sculpture by Barbara Hepworth.

A former music producer and professional photographer, Pitt gave meteorites a glamorous, *Pygmalion*-esque makeover. Rather than the forensic images familiar to the worlds of science and natural history, his photographs are flattering portraits reminiscent of both fashion and food photography. Dramatically lit, the meteorites hover in mid-air or are poised on minimalist pedestals against sateen backgrounds of carefully chosen colours. Close-ups of some meteorites' time- and space-weathered complexions emanate virility; the stony arabesques of polished slices emit a coquettish gleam and small amorphous masses pose as invitingly as boutique chocolates. Pitt also packaged meteorites, creating in 1997 the 'limited edition Mars Cube' a hermetically sealed vial containing a 1/10 carat morsel of the Zagami (Nigeria) Martian meteorite encased in a 6.35 cm Lucite block, an item that sold by the thousands. At an auction in 1998, a single gram of a Martian meteorite Pitt acquired from London's Natural History Museum, went for U.S.$16,000, over a thousand times the price of gold.[15]

Barbara Hepworth, *Single Form*, 1964, bronze sculpture 3.2 m high in Battersea Park, London.

The catalogue for the mega-meteorite sale hosted in 2012 by Dallas-based Heritage Auctions illustrates Pitt's talent for hitching the poetics of the *rarissime* to the luxury-goods bandwagon.[16] The descriptive copy accompanying the lushly photographed meteorites, many from the Macovich collection, is a marketing tour de force, wedding science and art to a patrician sensationalism and targeting a variety of sophisticated potential clients. A back story enhances a meteorite's desirability, the tales the proud owner can look forward to telling whenever someone asks what it is. The brief, instructive texts accompanying each auction lot offered something to suit every narrative propensity, all calibrated to appeal to a self-image marked by individualism, intelligence, artistic acumen and the irresistible urge to display surplus wealth.

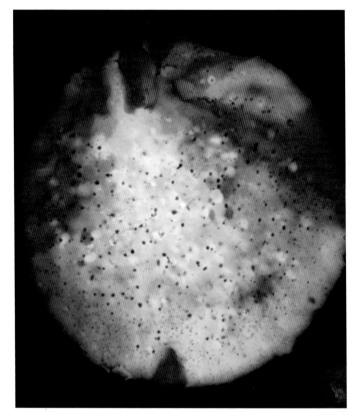

Thin section of the Kainsaz (Tatarstan) meteorite photographed in cross-polarized transmitted (pass-through) light at a magnification of 760×.

Rare specimen of the Ogi meteorite from an observed fall, 8 June 1741 on Kyushu, Japan. Portion of a mass held in the British Museum, London; 6.7 × 3.2 × 0.2 cm, 17.3 g.

A piece of Valera meteorite (Venezuela) found one morning beside a dead cow, 'the neck and clavicle of which had been pulverized', was auctioned with 'an official notarized affidavit' naming the meteorite as the cause of death. The Valera was further marked by 'a richly hued variegated matrix chock full of chondrules', the spherical silicate inclusions that differentiate meteorites from all earthly stones. A 119-g fragment of the Zunhua meteorite 'that punctured a Chinese farmhouse' in 2008 was expected to attract bids of U.S. $7,000–8,000. For the more romantic, a 17-g partial slice of a chondrite that fell in eighteenth-century Japan (valued at U.S. $4,500–5,500) was associated with the weaver goddess Shokujo and sequestered in shrines by Buddhist monks.[17]

Among the auction's offerings was a striking array of pallasites 'the most resplendent extraterrestrial material known to exist', including a complete slice of the Imilac (Chile) meteorite with its 'sparkling mosaic of crystalline olivine in a nickel-iron matrix'. The gem-like olivine is an extraterrestrial version of peridot, the birthstone for the month of August, as the catalogue helpfully notes for gift-shoppers. Another piece of Imilac 'from the highest driest desert in the world' is 'an animated specimen ... characterized by the ornate crenellations and toasted yellow pockets where olivine crystals were blasted out by the elements'. The Seymchan (Siberia) Sphere (500 g, 60 mm diameter), which calls to mind a ball of Italian nougat, was ground from a mass three times its size. 'Only after the needs of research institutions are addressed does the luxury exist to create [this] compelling presentation', valued at U.S.$4,000–5,000.[18]

'Scientifically exotic' specimens appeal to the connoisseur, such as the Krymka (Ukraine) meteorite, with its 'plethora of xenolithic clasts [some containing] *mysterite*, a finely grained dark material noted only in one other meteorite'. A slice of the Allende Meteorite (Mexico), a rare carbonaceous chondrite, holds traces of 'true stardust' (that is, calcium-aluminium inclusions) 'the first materials to condense out of the cooling nebular gases from which our solar system was formed'. Pedigreed meteorites placed on the block included bite-sized chunks of the l'Aigle meteorite from the 1803 shower that convinced the scientific community that stones did fall from the sky. A piece of the venerable Ensisheim (Alsace) meteorite was also on offer, whose fall in 1492 was construed as a favourable omen to the future Holy Roman Emperor Maximilian in his battle with France.[19]

Samples of the moon and Mars, the rarest and most costly meteorites, are in such demand for research and museum display that acquiring them for public sale was further testimony to Darryl Pitt's resourcefulness. From Mars, there was a 327-g chunk of the Tissint (Morocco) meteorite, 'blanketed in a glossy, glistening black fusion crust', a portion of the larger mass on display in London's Natural History Museum, and valued at U.S.$230,000–260,000. As for the moon, a scientific abstract from the *Meteoritical Bulletin* authenticated the origin of what the catalogue describes as 'the single most aesthetic lunar fragment

Mosaic of crystalline olivine in the nickel-iron matrix of the Imilac pallasite, from Chile's Atacama Desert. Only 1 per cent of all meteorites are pallasites. This specimen was cut from the centre of the main mass currently on display at the Natural History Museum, London; 46 × 46 × 0.2 cm, 2.699 kg.

on Earth'. Found by Berbers on the desert borders between Mali and Algeria, the lunar meteorite's 'otherworldly ornamentation' is owed to maskelynite, a shock glass which appears as 'an elegantly thin filigree of shock veins through the matrix' (103 g, bid range U.S.$55,000–70,000).[20]

Certain chemical and mineralogical properties add to a meteorite's beauty as well as its value, like the silvery geometries of Thomson Structures (or Widmanstätten patterns), octahedral cross-hatchings of kamacite and taenite that appear when the polished surface of iron meteorites is washed with a nitric acid

Precision-cut portion of the Muonionalusta meteorite showing Thomson Structures (Widmanstätten patterns). Found in 1906 in northern Sweden and estimated to have fallen around 1 million years ago; 9 × 9 × 9 cm, 5.66 kg.

solution. These nickel-iron alloys were formed as molten masses left over from the solar system's coalescence cooled in the vacuum of space, an infinitesimally slow process lasting millions of years. Some meteorites display striking regmaglypts, surface indentations formed when portions of the mass are scooped away by friction as it penetrates Earth's atmosphere. Also known as 'thumbprints', regmaglypts convey a sense of movement, like swirling tongues of flame cast in iron.

The round holes or deep voids left in some meteorites from the disintegration of spherical inclusions (more fragile than the matrix) are also considered highly aesthetic. The most valuable sculptural pieces of the Heritage Auction in 2012 were Gibeon irons from a copious prehistoric fall in Namibia. An 81-kg rectangular meteorite with three large asymmetrical voids that

Oriented Henbury meteorite with marked regmaglypts, from the Henbury Crater field in Australia's Central Desert.

double for eyes and a gaping mouth was entitled 'the otherworldly scream' for its resemblance to Edvard Munch's famous painting. The auction's centrepiece was the owlish 'Gibeon Mask', its pair of deep-set holes flanking a flinty ridge in a 'singular zoomorphic evocation' (bid estimate U.S.$140,000–180,000). Big is also beautiful, as demonstrated by the Nantan meteorite, a 750-kg 'extraterrestrial colossus' found in Guangxi Province, China, and mounted on a 'custom armature' (U.S.$85–115,000).

The Gibeon strewn field in Namibia yielded many aesthetic meteorites, including this 'otherworldly scream'; 55 × 24 × 21 cm, 81.37 kg.

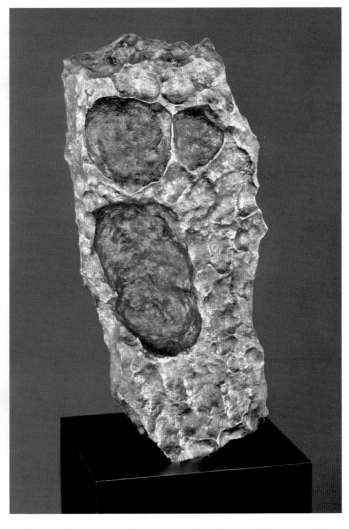

'Gibeon Mask' found in Africa's Kalahari Desert, 1992, a singular natural sculpture from outer space; 20 × 21 × 18 cm, 9.37 kg.

Likewise, 'if size is what matters', as the catalogue candidly suggests, the Mundrabilla (Australia) meteorite slab 'is not a tabletop specimen – it's the table', weighing 721 kg (bid estimate u.s.$85,000–115,000).[21]

Meteorite collectors are unperturbed by the contradictions that may arise when the objects of their fancy are presented as equally valuable to science, art and interior design. To them, this is just good PR; meteorites proclaiming their myriad qualities to the world. Aficionados and scientists alike point out that high-profile auctions have inspired legions of new hobbyists to hunt for meteorites, providing specimens for research that might otherwise have been lost to 'terrestrialization', the process by which meteorites lose their cosmic chemistries to weathering, water and rust. According to Darryl Pitt, 44 Martian, 46 lunar and other exotic meteorites have been recovered from desert falls since the mid-1990s by local agents and all were eventually made available to researchers. Scientists themselves have only gathered four such specimens in the same period.[22] Carl Agee, director of the Institute of Meteoritics in New Mexico, agrees that 'the number of rare Angrite meteorites has more than doubled due to African finds – a huge enhancement to our understanding of the early solar system.'[23] Owing to the difficulty researchers would have recovering meteorites from remote locations around the world, it may be argued that the private sector (hobbyists, traders and/or their agents) has proved indispensable to the advance of meteoritics.

Classified meteorites are more valuable than unclassified ones and reputable collectors and traders deal only in specimens authenticated by the Nomenclature Committee of the Meteoritical Society.[24] Many collectors feel a 'social responsibility' to give scientists 'a first crack' at examining newly acquired rare specimens.[25] Researchers may also obtain pristine finds collected during the annual NASA-funded ANSMET (Antarctic Search for Meteorites) missions to Antarctica which are shipped, still frozen, to the Johnson Space Center in Houston, Texas, where thin slices are prepared for distribution to researchers worldwide, free of charge. Since 1976 ANSMET has recovered over 20,000 meteorites. Pitt correctly asserts that the sale of sculptural iron meteorites in no way 'compromises science', especially since those he places on offer are part of larger falls that amply fill researchers' needs. But the commoditization of meteorites has had unintended consequences that cannot be so readily dismissed.

While raising the public's awareness of meteorites, auctions resulting in major sales drive up prices; acquiring important meteorites for public display costs money that museums and

Meteorite find at Mount Pratt, January 2005, U.S. Antarctic Search for Meteorites Program.

educational institutions cannot always afford. The Museum of Natural History in Vienna paid a German dealer over $500,000 (€400,000) for a 950-g piece of the Tissint Martian meteorite in 2011, now a key feature of their recently revamped permanent display. When asked if the price seemed a bit steep, Dr Franz Brandstätter, curator of the meteorite collection, shrugged and said '[some] collectors can reach into their pockets for a million dollars; we were happy to get it at all.'[26]

Another spin-off of the growing demands of hobbyists and traders is that meteorite hunting has become a profitable pursuit in relatively impoverished places such as Namibia, where, Pitt notes, 'local tribesmen used [Gibeon] iron meteorites as spearpoints 100 years ago. In the 1990s their [descendants] used metal detectors to look for iron meteorites to sell.'[27] Northwest Africa is another treasure trove for hunters and the savvy traders that keep them in business. Working in partnership with a Moroccan dealer and 'nomad' hunters, the Canadian trader Dean Bessey claims to have 'gone through tons of meteorites and supported the Saharan economy to the tune of around a half-million dollars' over the course of three years.[28] The Saharan 'informal economy' might have been a fairer assessment, unless nomads file tax returns. In addition, not all the specimens entering the market from far-flung places are officially classified.

According to ANSMET geologist Ralph Harvey, 'the mass removal of meteorites from [desert] countries led to many realizing what they were losing, which subsequently led to some cracking down on meteorite hunting.'[29] Yet even in countries where laws exist to regulate meteorite export, enforcing them is another matter. In 2011, following the discovery of a 5,000-year-old crater and strewn field deep in Egypt's Western Desert, Gebel Kamil meteorites began surfacing in online sales catalogues and at gem and rock shows. Likewise, within days of the copious fall in Chelyabinsk in 2013, a Russian website dedicated to classified advertisements overflowed with offers of meteorites ranging in price from u.s.$20 to u.s.$3,300. Some thoughtful individuals offered 'a private tour of the crash site and sightseeing excursion of the destruction' for u.s.$167, including airport pickup.[30] Locals

Gebel Kamil meteorite lying *in situ*, Egypt, 2010.

sold their finds to mysterious out-of-towners, men who 'refused to answer questions' but had plenty of ready cash.[31]

That meteorites should be subject to the same capitalist imperative as other precious substances is neither new nor surprising. Following a shower of meteorites in Siena in 1794, locals did a brisk business selling them to English tourists and when they ran out, the ordinary stones they substituted were presumably as well-received.[32] The practice of such relatively harmless fakery has been known ever since enterprising residents of ancient Thebes furnished Greek and Roman tourists with faux scarabs indistinguishable from the genuine, thousand-year-older item.[33] While meteorites, unlike antiquities, may be sliced and diced to augment the supply, they are still difficult to come by and more easily faked, especially when sold online. Buyers may order a rare specimen but receive less valuable substitutes, and unless it is analysed they will never know they have been cheated. 'Authenticity' is the motto of the International Meteorite Collectors Association, a group of self-monitoring amateur and professional collectors formed to combat meteorite scams, and there are several websites offering consumers tips on how to spot the obvious fakes.[34] At least one man has been convicted of meteorite fraud in a court of law. Steve Curry of Colorado offered phonies for as much as u.s.$500,000 apiece and threatened museum curators who refused to authenticate them.[35]

Theft is another consequence of meteorites' highly publicized price tags. A heist at a science education centre in North Carolina on Christmas Eve 2012 resulted in the swift apprehension of a 29-year-old man who was so intent on making off with 100 meteorites and computer equipment that he ignored another hot electronic item, the surveillance camera.[36] Meteorites are occasionally snatched at trade shows and exhibits. Some of the specimens are too well known to be resold. Those willing to steal to simply possess them are not ordinary thieves, writes Kevin Kichinka, author of *The Art of Collecting Meteorites*, but 'victims of the singular and clinically unrecognized power of a meteorite over a person'.[37]

Meteorite hunting is an open-air sport pitting collectors against both the elements and the odds. It attracts a certain kind of individual and may translate into a lucrative, alluring profession that in many ways recalls the historic pursuit of treasure hunting. Consider 'the Seekers', a medieval guild in Egypt that paid the caliph a tax in order to dig for ancient gold. It was a dangerous livelihood requiring not only physical prowess and artisanal skill but also the courage and perseverance to tunnel into the earth and penetrate stone-walled tombs said to be guarded by fierce *djinn* (genies). The *One Thousand and One Nights* is full of occult-tinged tales of treasure hunters, whose profession was so attractive that popular how-to-hunt manuals were compiled, replete with advice on subduing noisome djinn. In the fourteenth century amateur treasure hunting was common enough to be considered a social ill akin to gambling. As archaeologists began to assess the value of ancient artefacts in the mid-nineteenth century, travellers flocked to Egypt's ruins. Acquiring ancient trophies, whether personally or through some obliging local, was the prime objective of many a journey.[38]

Today's professional meteorite hunters, who call themselves 'meteorite men', pay taxes to their governments.[39] Accustomed to roughing it in all sorts of terrains, they possess an understanding of the pertinent branches of science (geology, mineralogy, astronomy and so on) and nurture a profound respect for the forces of nature. Global positioning systems guide them to their hunting grounds, the strewn fields of our planet, and their metal detectors make the dumb earth speak, revealing what lies beneath as if by magic. Stars of a thousand and one television documentaries, authors of scores of books, articles and blogs, their exploits have attracted bands of followers worldwide, inspiring what has been called a 'twenty-first-century gold rush' for meteorites.[40] Only Antarctica, claimed by the ANSMET programme, is entirely off limits to hunters; the rest of the world is more or less theirs.

Professional meteorite hunters come from a variety of backgrounds but have much in common, including a fascination with outer space often conceived as kids watching Neil Armstrong's moonwalk or having witnessed an exploding fireball. They are non-conformists, fans of the wondrous; launching their careers usually involved selling things, starting over, pinning their fortunes literally to a star. Using the sale of some finds to finance hunts for others, they are happiest when supplying rare specimens for connoisseurs. As such, they all stand on the unassuming shoulders of Harvey Harlow Nininger (1887–1986), the bespectacled tutelary deity of meteorite enthusiasts great and small. A real-life candidate for G. K. Chesterton's fictional *Club of Queer Trades,* it was Nininger who invented this wholly new occupation – finding, collecting, trading, selling, studying and exhibiting meteorites – and made his living from it exclusively, albeit barely.

Meteorites were more than a living to Nininger: they were a calling he answered with heroic perseverance throughout a turbulent century. He understood that the science of meteoritics could only be advanced if new specimens were found, distributed and studied. In his memoir, *Find a Falling Star* (1973), he wrote:

> It was a source of some chagrin to be introduced, as I was frequently, as 'the man who has found more meteorites than any other man in history.' Such a statement missed the main point of my life. Collecting occupied much of my time and effort, but collecting served as a sort of platform or footing on which to stand while I sought to educate, and while I pleaded constantly for an organized program of meteoritical research.[41]

Nininger was preoccupied with education for having felt the lack of it. By the age of fourteen, at the start of the twentieth century, his family had moved several times, from Kansas to Missouri and Oklahoma, travelling in covered wagons and on stock trains. He attended school for the three months of the year when he wasn't picking cotton, and was expected to become a farmer. Two straight years in an Oklahoma high school awakened his interest in science, which his family feared would 'spoil his religion'. In 1909, against their wishes, he began studying biology at MacPherson College, Kansas, where he was invited to teach while still completing his graduate degree. During the First World War he worked for the u.s. State Department's

Harvey Harlow Nininger wielding his homemade 'metal detector' (a magnet attached to a stick), photographed near Meteor Crater by his wife Addie, late 1940s– early 1950s.

Portion of a chondrule in a thin section of NWA 5142, a stony meteorite containing very little iron, photographed in cross-polarized light at a magnification of 160×.

food conservation programme, and was posted to South Dakota to study crop predators such as grasshoppers and locusts. Nininger had never heard of meteorites until reading an article in *Scientific American* in 1923, by which time he was the head of MacPherson's biology department and a married man with three children. Several weeks later he witnessed a fireball explode high above the Kansas plains and his life's direction changed. 'I was cocked and primed and started right in to chase that fireball. Been at it ever since!' he said in an interview in 1976.[42]

Watching the sky that November night, Nininger was certain that fragments of the exploding body had reached earth near the tiny town of Coldwater, Kansas, where he subsequently placed advertisements in papers and canvassed schools and churches, offering a dollar a pound for recovered fragments. Instead, farmers brought him odd stones they'd held on to for years, some of which turned out to be 'cold finds', meteorites from unwitnessed falls. Nininger realized that America's Midwest was a fine place to search for meteorites; its vast tracts of farmland had long ago been cleared of rocks and if annual ploughings upturned new ones, some were probably extraterrestrial. His approach was to educate people about what to look for and why. From 1923 to 1929 he spent his weekends driving hundreds of miles from home, going farm to farm, lecturing, writing articles about meteorites and distributing leaflets. Sometimes he would hear from people many years after his visit; they never forgot him and posted him samples of strange rocks they had found, some of which were meteorites.

Convinced that meteorites were more abundant on the ground than had been previously thought, Nininger yearned for more time and funds to enlist the public in finding them. Dr George P. Merrill, geologist and curator of the Smithsonian Institute's meteorite collection, responded to his request for assistance with scepticism: 'Young man, if we gave you all the money your program required and you spent your life doing what you propose, you might find one meteorite.' Nininger rode out the Great Depression in Denver, working at the Colorado Museum of Natural History and writing his first book, *Our Stone-pelted Planet* (1933). During the Second World War he joined the war effort, searching for scrap iron in mining towns. By the time the war ended, with the help of his Midwestern network, Nininger had found over 5,000 meteorites from 526 falls, the greatest personal collection ever assembled, and with his wife Addie decided to start a museum of their own.[43]

Opened in 1946, the American Meteorite Museum was located on Route 66 in Arizona near Meteor Crater, where Nininger foraged for meteorites and explanations of the crater's formation. Admission to the museum cost 25 cents and for a while tourists came by the thousands, as did researchers, who purchased specimens for educational and scientific institutions worldwide. When Route 66 was shut down and traffic diverted, the museum was left marooned. Nininger struggled on for seven years amid repeated refusals from private and state agencies to his petitions for financial assistance. Sales of trinkets fashioned from small oxidized fragments helped the family survive, at the cost of criticism from an uncharitable academic community that shunned Nininger for commercializing meteorites and belittled his contributions to meteoritics.[44]

In 1953 the family managed to move their collection to Sedona, Arizona, quite a task considering its size and weight. Unlike the old museum, this one had electricity, but visitors were too few to pay the bills. Nininger reluctantly sold a fifth of his collection to the British Museum for U.S.$140,000 in 1958. In 1960 Arizona State University purchased the rest for U.S.$275,000, fulfilling Nininger's wish that the meteorites remain in the United

States and become the focus for research, in this case Arizona's
Center for Meteorite Studies. His family was finally solvent, able
to build a home and travel the world, always looking for meteorites.
Nininger, who had authored ten books and over 150 papers,
continued lecturing well into his nineties.[45]

Dr Frederick C. Leonard (1896–1960), professor of astronomy at the University of California at Los Angeles, shared
Nininger's conviction that people should know more about
meteorites, and was one of the first to teach a course entirely
devoted to them in the United States. Together they founded the
Society for Research on Meteorites in 1933, a time when
America had no research programme and meteorites were
scarcely mentioned in high school or college curricula. Leonard
took his classes to Nininger's museum and on meteorite hunting
expeditions around Meteor Crater. The experience made several
of his students lifelong devotees, including O. Richard Norton
(1937–2009). Professor of astronomy and space sciences for over
40 years, Norton authored the illustrated *Cambridge Encyclopedia of Meteorites* (2002) and *Rocks from Space* (1994), both
invaluable field guides and holy writ to meteorite hunters and
collectors. With his wife and collaborator, illustrator Dorothy
S. Norton, Norton welcomed enthusiasts into their home and
became a kind of guru. It was he who handed a young Robert
Haag, now a renowned 'meteorite man', the first space rock he

ever touched. 'The air was electric', Norton recalls, as Haag examined the stone: 'here was a man possessed'.[46]

The son of commercial rock and mineral dealers from Tucson, Arizona, Haag had witnessed an exploding fireball in Sonora Mexico at the age of thirteen and heard one of Nininger's last lectures. 'Space', writes Norton, 'occupied his waking hours'. Norton first encountered Haag in a Tucson shopping mall; 'a wild looking fellow in a silver jumpsuit' standing beside an huge picture of the moon selling 'space passports' (a snapshot of the prospective space traveller mounted in a pocket-sized cardboard moon) for $5 apiece. Haag believed that commercial space travel was in the offing and couldn't wait to go. He had the 'look of a rock star and the smile of a televangelist', according to Norton, who figured Haag for a con artist. Norton met him properly, having answered an advert in a local paper looking for meteorites, without knowing Haag had placed it. Haag arrived with his arm in a sling from a hang-gliding accident. The two were soon fast friends.[47]

With Norton as his mentor, Haag became an avid meteorite hunter with important finds to his credit, including a quantity of the Murchison (Australia) meteorites found in the early 1980s.[48] These were in demand ever since samples from the fall of 1969 were found to contain a number of amino acids, the building blocks of life. The supply of fresh samples was limited, and meteorites of this type deteriorate rapidly owing to their un- usually high water content of 10–14 per cent. Although the 1969 shower had covered 33 sq. km, there was little hope that portions had survived. Instead of searching the ground, Haag followed Nininger's example, canvassing the small town of Murchison, enlisting its inhabitants and eventually knocking on the right door. Mrs Betty Maslin had witnessed the fall and recalled the loud explosions, hissing noises and strong smell of methyl alco- hol that filled the air that September day. She sold Haag a jarful of 'the smelly rocks' she had collected and kept carefully sealed for over ten years. 'When I opened it,' wrote Haag,' the smell of alcohol and ether was still strong enough to nearly put me under.[49] The Murchison find established Haag's reputation as someone who could deliver precious specimens. Far from begrudging his protégé's success, Norton defended Haag's enterprise, pointing to his contributions to the pool of specimens available for study. As the first private citizen to obtain both lunar and Martian meteorites, Haag became a minor celebrity, making headlines and frequent television appearances. He never took the trip to space as he had wished, but the man Darryl Pitt called 'the P. T. Barnum of meteorites' had travelled a long way since the days of the silver jumpsuit.[50]

Showmanship is not uncommon among meteorite men. Geoffrey Notkin, author of an engaging memoir entitled *Rock Star: Adventures of a Meteorite Man* (2012), started out as a punk rock musician. Notkin recounts his transformation from a leather-clad denizen of the Lower East Side punk scene to intrepid outdoorsman, trekking across America, Chile's Atacama Desert, the Australian Outback, Iceland and Siberia in search of meteorites. He recalls fossil hunting with his mother and, as a boy of eight, filling holes in the beach with bits of burnt wood, pretending they were impact craters. The punk rock revolution 'neatly sublimated' his boyhood interest in geology and meteorites, but reading Norton's *Rocks from Space* in 1999 revived 'all the wonder and anticipation' he had felt as a child. Notkin dusted off the rock hammer he had always kept in the trunk of his car, purchased some maps and a compass and started hunting.[51]

Aside from surviving earthquakes and hurricanes, being stranded in the desert and having 'a finger crushed between two shockingly powerful rare earth magnets', Notkin inventories the perils of his profession:

> my exterior surface [has been] scratched, cut, burned or otherwise injured by barbed wire, rusty vintage farm machinery, a portable gasoline stove . . . poison ivy, scorpion weed, other unidentified allergens, rocks, rock hammers, bamboo, cornstalks, every kind of thorn, bramble and cactus, and . . . I have been bitten by mosquitoes, gnats, chiggers, black flies, robber flies, fire ants, various types of spider and a cattle dog . . . [had] close encounters with rattlesnakes . . . dog-sized

lizards with blue tongues, alligators, wild boar, tarantulas, black widow spiders, two swarms of killer bees . . . scorpions, the venomous giant desert centipede (*Scolopendra heros*), two angry bald eagles with very large talons, the Sonoran lynx (*Lynx rufus*) and more than one pack of hungry coyote.[52]

Notkin made significant finds with Kansas-born Steve Arnold, whom he met online. An experienced meteorite hunter, Arnold devised a trolley for hauling a powerful but cumbersome metal detector, 'a versatile, lightweight rolling vehicle the size of a large coffee table, rough enough to be towed behind a motorcycle across farmland for ten hours a day'. And he built it, Notkin notes admiringly, with 'no screws, no wire, no washers, no ball-bearings [since] even one metallic component would cause continuous activation of the coil, rendering the detector useless'.[53] Notkin and Arnold appeared in several documentaries before landing their own three-season show, *Meteorite Men*, on the popular American Science Channel, featuring their hi-tech hunts in remote locations. Together they tracked down meteorites from fresh falls and forgotten ones, camping in craters and travelling with portable mini-tractors and a Hydratek ('a go-anywhere vehicle') nicknamed 'Rock-hound'.

Owner of Aerolite Meteorites (whose motto is 'we dig space rocks'), Notkin provides specimens to researchers, institutions and collectors worldwide. Not adverse to publicity, he has appeared in adverts for Fischer Research Laboratory, Texas, the people who patented metal detectors and manufacture a variety of these instruments, which resemble an orthopaedic crutch with an oval plate at the bottom. An award-winning science writer, Notkin tirelessly sings the praises of the space rocks that have so variously enriched his life. 'Like the sound of the ocean in a seashell, meteorites carry within them a faint murmur of infinity', he writes. But his biggest thrill was having an asteroid named after him; 'minor planet 132904', discovered at Palomar Observatory by his friend the astrophysicist Robert Matson, is now officially 'Notkin'. In his memoir Notkin dedicates the honour to his father: 'Somewhere up there in the maelstrom of asteroidal debris between Mars and Jupiter, is a somber little world with Dad's name on it.'[54]

Although meteorite hunting has risen in some circles to a profitable vocation somewhere between performance art and quest for enlightenment, it is nonetheless a rote endeavour demanding dogged persistence. In *Rocks from Space*, Norton offers crude statistics suggesting that 100 meteorites may be found on each 2.5 sq. km (1 sq. mi.) of ground based on an estimate of their accumulation during a million-year span and supposing they were evenly distributed. If only 10 per cent of these could be found, that would be ten per square mile, ten times what Nininger had calculated in his day.[55] Wide-open spaces where meteorites would have fallen and accumulated over great stretches of time are ripe for hunting. But even when searching in an identified strewn field, it can take a lot of work. Consider the travails of Steve Schoner, who decided to search for pieces of a pallasite that fell on a ranch near Glorieta, New Mexico, in 1884. Although 174 kg had been found, there was reason to believe more remained. Schoner combed the strewn field for weeks at a time during 70 trips, each 1,400 km by car from his home, and was finally rewarded for his efforts with a splendid 20-kg hunk of Glorieta in 1997, fifteen years after he started looking for it.[56]

Complete slice of the Glorieta Mountain pallasite featuring signature dark-grey inclusions of iron sulphide along with an aggregate of olivine at the right margin with coursing tendrils of nickel iron; 12 × 15 × 0.1 cm, 191 g.

Meteorite hunter Ralph 'Sonny' Clary of Las Vegas, Nevada, found a way to improve his chances with the help of an accomplice: a German Shepherd named Brix. A firefighter by profession, Clary adapted techniques used for training cadaver dogs to teach Brix to sniff out freshly fallen stony meteorites. Brix was additionally taught to avoid snakes, a lethal hazard in the deserts and dry lakebeds of Nevada where Clary hunts, using a Pavlovian technique involving caged rattlers and mild electric shock. Accustomed to rushing to the scene of conflagrations, Clary's particular interest is investigating fireball sightings, sometimes with the assistance of Doppler Radar technology to track the trajectories of exploding fragments and pinpoint strewn fields. He also consults the Meteoritical Society's online bulletin for the locations of previous falls, since the best places to hunt meteorites are where they have already been found.[57] With discoveries numbering in the thousands, Clary has donated kilograms of meteorites to educational outreach programmes in the USA and UK, in addition to samples of fresh finds to universities and the Smithsonian Institute.

One of Clary's hunting companions is fellow Nevadan Count Guido Deiro, a pilot, champion fencer and former manager of the Las Vegas Airport, which was acquired by Howard Hughes in the 1960s. Deiro piloted for Hughes, on one occasion delivering him to the Cotton Tail Ranch, a renowned Nevada brothel where he claimed to be searching for 'a red-headed hooker with a diamond in one tooth'.[58] Deiro, who would later spearhead the development of the Las Vegas Speedway, inherited his title from his father, who was bestowed it in his native Italy for his achievements as a composer and accordionist. The indomitable Count Deiro *fils* started meteorite hunting at age 72 as a means of restoring his health and well-being after being diagnosed with cancer. On his second hunting trip with Clary and Brix, he discovered his first meteorite, the largest (12.7 kg) chondrite ever found in Nevada.[59]

Deiro found his prize while waving a staff with a neodymium magnet taped on the end at the ground, a homemade device that demands concentration; there are no flashing lights,

Sonny Clary's meteorite-hunting German Shepherd, Brix, and his find, the 205-g Mifflin, L5 chondrite, April 2010, Wisconsin.

beeps or clicks when it approaches metal-containing objects, it is merely drawn to them, sometimes almost imperceptibly. And not all meteorites contain large amounts of iron; the rarest ones (lunar, Martian and carbonaceous chondrites) may contain none at all. The experienced meteorite hunter is familiar with all types of meteorites and how they weather with exposure. A sharp eye and knowledge of the terrain is essential, but it takes almost a sixth sense to notice meteorites lying among terrestrial rocks. Sonny Clary, who seems to have the knack, prefers 'cold hunting' to using a metal detector; he just picks an open, undeveloped stretch of land and starts walking, his eyes glued to the ground. As Clary points out, meteorites have been falling for millions of years, and if you look hard enough you're bound to find one. 'I just enjoy the outdoors and knowing meteorites are everywhere', he says.[60]

Hobbyists are advised to carry a magnifying glass (10× power, like a jeweller's) to study the surface of stones, and a diamond sharpening stone to open 'windows' to examine the interior for the tell-tale signs of cosmic origin, including chondrules, olivine crystals or metallic grains. Kevin Kichinka's *The Art of Collecting Meteorites* offers tips for cataloguing and caring for specimens, noting that 'for rocks from space, Earth's environment is the harshest of all'. The first to find a meteorite in Bolivia, Kichinka divides his time between meteorites, guiding rainforest and volcano treks, classical piano and meditation. Not all collectors can afford museum-quality conditions for their meteorites – climate-controlled, hermetically sealed, bulletproof glass cabinets – but according to Kichinka, some can and do. Everyone else should keep their meteorites as dry and clean as possible; they are prone to rust and Kichinka recommends lubricating them with gun oil to slow the process. Dipping meteorites in paraffin helps preserve them but leaves a matt finish.[61] Wearing gloves during these ministrations is a must, since hands carry corrosive oils and salts. Some collectors bake their specimens to drive off humidity but a little pouch of desiccant also helps. Sometimes all this loving care is in vain. Not only can fondling your meteorites eventually ruin them, but some iron-bearing ones are susceptible to Lawrencite disease, a meteorite cancer that crumples them to dust.[62]

A perusal of print and online literature regarding meteorites is further proof of the fascination they hold.[63] Publications like *Meteorite Magazine: The International Quarterly of Meteorites and Meteorite Science* and *Meteorite Times.com* feature contributions from both scientists and aficionados, such as former chemist Robert Warin and former civil engineer John Kashuba, who have devoted their retirement to analysing the mineralogical complexities of chondrules and the carbon-based molecules they contain. Or Tom Phillips, another dedicated hobbyist whose 'micrographs' (photographs of thin sections under a high-powered microscope) illustrating the infinite variety of meteorites' internal structures (magnified up to 1,600 times) stand at the intersection of art and science.[64] The editors of these specialist publications typically have a science background and often double as contributors, a labour of love performed in their spare time.[65] It's as if meteorites had generated a sort of character: outdoorsy, gentlemen scholar-technicians with a hint of the temple priest. In *Consciousness Explained* (1991) Daniel C. Dennett suggested that 'a scholar is just a library's way of making another library'. Perhaps a meteorite collector is just a meteorite's way of securing another collection.

Among the most unusually shaped meteorites known, this Gibeon iron was formed in part by terrestrialization (slow oxidation on Earth over thousands of years); 80 × 28 × 20 cm, 61 kg.

And why not? Meteorite collecting offers wholesome exertion in the arms of nature alongside the intellectual stimulation and emotional thrills of the great unknown. The finds are nothing less than cosmic realia, mementos of a life virtually untouched by human experience. The narrator of G. K. Chesterton's *The Club of Queer Trades*, surveying his ten fellow members, expressed the same sentiment that seems to drive all meteorite aficionados:

> To realize there were new trades in the world was like looking at the first ship or the first plough. It made a man feel what he should feel, that he was still in the childhood of the world.[66]

Here, finally, is where meteorites differ from antiquities. Although infinitely older, they evoke not the past but the future, something temporarily beyond our grasp yet still within reach. And in the place of appreciation for the achievements of bygone civilizations, meteorites inspire a sense of striving for discovery, the ever-new.

4 All Things Said and Done

The world was meant to end in 1492, according to some interpretations of the Bible's Revelations to John, and the deafening explosion that preceded a large meteorite fall near Ensisheim, Alsace, on 7 November that year must have struck many as the wrath of God. Resounding over an area of 40,000 sq. km, 'there fell a burning stone', wrote the German poet Sebastian Brant (1457–1521), who was near Ensisheim at the time:

> Singed and earthy and metalliferous . . .
> The explosion was heard on both sides of the Rhine
> Heard also by the Uri among the Alps,
> It astounded the Noricians, the Swabians and the Rheticans:
> It sounded in the Burgundian ears, and caused the French
> to tremble.[1]

It was the first witnessed fall in Europe since the invention of the printing press, and Brant's poem appeared on four widely distributed broadsheets within several weeks. The topical verse, written in both Latin and the vernacular German Brant wished to promote, was accompanied by woodcuts of a contused sky rent by blades of light, with a murky mass plummeting towards the walled city.[2]

If Brant hoped to expand his readership he had a particular reader in mind, Maximilian (1459–1519), son of the Holy Roman Emperor Frederick III, who was leading an army towards Ensisheim en route to battle France when the meteorite fell.

Thin section of the Juancheng H5 chondrite, from a copious meteorite shower in Shandong, China, 15 February 1997, photographed in cross-polarized light at a magnification of 160×.

The only surviving original of Sebastian Brant's first broadsheet describing the fall of the 'Donnerstein' at Ensisheim in 1492. The Latin and German verses describing the fall are followed by an address to Holy Roman Emperor Maximilian.

Brant's poem concluded with a dedication to the king proclaiming the fall an omen of his imminent victory:

> Take as truth the stone was sent to you,
> God warns you in your own land
> That you should arm yourself.
> Oh mild King, lead out your army
> Let armour clang and roar of guns,
> Let triumph resound;

Slice of the Ensisheim meteorite, 3.1 × 2.4 × 0.2 cm.

Curb the swollen pride of France
Preserve your honor and your good name.[3]

Arriving in Ensisheim, Maximilian hacked off a piece of the 135-kg stone as a memento of the 'miraculous event' and left the rest in the church of Ensisheim for safekeeping, where a portion remains to this day. When the king effectively trounced the French at the Battle of Salins in January 1493 despite having far fewer troops, Brant could not resist reminding the public of his prescience in another broadsheet, while laying it on thick for Maximilian:

> As I told you earlier,
> The stone does not lie . . .
> The luck which it brings you this year
> Will follow you and
> Be true to you until you leave this life.[4]

Albrecht Dürer,
A Heavenly Body,
c. 1496, believed to
depict the Ensisheim
fall of 7 November 1492.

Brant milked the meteorite yet again when Frederick III died in August 1493, suggesting it also foretold Maximilian's succession. Call it overkill, but these sensational, journalistic broadsheets placed Brant squarely in the public eye, which probably helped his social satire *Ship of Fools* (1494) run to numerous official and pirated editions; woodcut illustrations by a young Albrecht Dürer (1471–1528), his first commissioned work, enhanced its appeal to the literate and illiterate alike. Dürer was coincidently near Ensisheim in 1492, and a peculiar painting of a brooding sky streaked with lurid flames appears to be his rendition of the event. The unsigned oil was found in the 1960s on the back of a small wooden panel depicting St Jerome and attributed to Dürer.[5]

Whether or not Brant actually believed in the Ensisheim meteorite's oracular or talismanic properties, he knew his audience did: his poem cited the portents and prodigies recorded by the venerable Pliny ('milk raining from the sky, grains of steel/And iron, flesh, wool and gore'), assigning Ensisheim a place in a long-accepted tradition.[6] And so it was that meteorites made their print debut, with Brant's opportunist verse

A depiction in ink and
tempera on parchment
of the explosion of the
Ensisheim fireball and
the fall of the stone into
a field. In place of the
boy who was the sole
witness to the fall, this
fanciful scene shows a
field being sown by two
men with a large horse.

ploughing the first fertile furrows in a field of narratives. Rooted in myth and historic annals, seeped in apocalypse and redemption, meteorites would soon yield a cornucopia of literary and artistic fruits.[7]

When William Shakespeare (1564–1616) began composing his works, the planets were held to guide man's fate and the universe believed to revolve around Earth. The hundreds of references to astral bodies in Shakespeare's works reflect both his personal interest and their familiar place in contemporary life; people used sundials, sometimes pocket ones, to tell the time and it was not uncommon to gauge the hour of the night by the position of the constellations. The movement of the planets and stars served Shakespeare as similes, as when Juliet (Act II, Scene 2) entreats Romeo, 'O, swear not by the moon, th'inconstant moon . . . lest they love prove likewise variable' or when Julius Caesar (Act III, Scene 1) tells Brutus he is 'constant as the northern star / Of whose true-fixed and resting quality / There is no fellow in the firmament.'[8] Shakespeare sometimes called Earth 'the centre', in keeping with the prevailing belief in its fixed position, and employed it as a simile for certainty and steadfastness; 'For my grief is so great that no supporter but the huge firm earth can hold it up' (*King John*, Act III, Scene 1). Comets were instead unruly and boded no good, as in *Henry VI Part I* (Act I, Scene 1):[9]

> Comets importing change of times and states,
> Brandish your crystal tresses in the sky,
> And with them scourge the bad revolting stars
> That have consented unto Henry's death!

Meteors were 'a prodigy of fear' and 'a portent / of broached mischief to the unborn times' *Henry VI Part I* (Act I, Scene 5). Shakespeare makes at least one dark reference to meteorites ('thunder-stones' in contemporary parlance) as two boys sing a funeral song over the body of a companion in *Cymbeline* (Act IV, Scene 2):

Fear no more the lightning-flash,
Nor the all-dreaded thunder-stone;
Fear no slander, censure rash;
Thou hast finished joy and moan . . .

Although Shakespeare used comets and meteors as dramatic devices alongside soothsayers' prophecies, he was the contemporary of astronomers who dared challenge geocentrism and with it, religion and superstition. Tycho Brahe, Giordano Bruno and Galileo advanced the Copernican theory of heliocentrism along with Johannes Kepler, Brahe's assistant, who redefined the laws of planetary motion.[10] As a lucrative sideline, Kepler made astrological forecasts for royalty, while openly disavowing belief in his own predictions.[11] Likewise, Shakespeare distinguished the popular meaning of astronomy (astrological prognostication) from the emerging, scientific one: 'Not from the stars do I my judgment pluck; And yet methinks I have astronomy' he writes in Sonnet xiv.[12] Edmund in *King Lear* (Act i, Scene 2) seems to express Shakespeare's own opinion:

This is the excellent foppery of the world, that, when we are sick in fortune, – often the surfeit of our own behaviour, – we make guilty of our disasters the sun, the moon, and the stars: as if we were villains by necessity; fools by heavenly compulsion . . .

Julius Caesar (Act i, Scene 2) famously remarked: 'The fault dear Brutus lies not in our stars / But in ourselves, that we are underlings.' And in *King John* (Act iii, Scene 4), Shakespeare decries the power of ignorance:

No scope of nature, no distemper'd day,
No common wind, no customed event,
But they will puck away his natural cause
And call them meteors, prodigies and signs,
Abortives, presages, and tongues of heaven,
Plainly denouncing vengeance upon John.

Social commentary aside, most of Shakespeare's references to the heavens aimed to convey a sense of beauty and majesty, 'a veritable sky psalmody' heightening the tone of the drama.[13] His 'bits of true or fanciful astronomical lore [made] facets of gem-like brilliancy shine out from his verse', wrote astrophysicist C. G. Abbot in 1938; 'what added beauties would he have created had he possessed the knowledge of the universe that awes us now, it is difficult to imagine'.[14] One also wonders what Shakespeare would have thought had he known that a fragment from *Romeo and Juliet* (Act III, Scene 2) would be chosen as the epitaph accompanying the ashes of Eugene Shoemaker (1928–1997), pioneering astrogeologist, on a NASA launch to the Moon:

> . . . and, when he shall die,
> Take him and cut him out in little stars,
> And he will make the face of heaven so fine
> That all the world will be in love with night,
> And pay no worship to the garish sun.[15]

These lines were laser-etched on piece of brass foil wrapped around the vial containing Shoemaker's ashes and, in a nod to the old beliefs, illustrated with the Hale-Bopp comet that blazed though the skies the year that Shoemaker died.[16]

The certainty of Earth's fixed place in the solar system slowly began to shift in Shakespeare's lifetime, when Galileo's telescopic observations showed our planet did indeed move. The new astronomy threatened to literally turn the world upside down and John Donne, writing in 1611, betrayed some perplexity:

> And new Philosophy calls all in doubt,
> The Element of fire is quite put out;
> The Sun is lost, and th'earth, and no man's wit
> Can well direct him where to look for it.[17]

The artist brothers Paul and Thomas Sandby witnessed the fragmenting fireball of 'The meteor of Aug 18th, 1783, as it appeared from the NE Terrace, at Windsor Castle' along with the physicist Tiberius Cavallo, who documented it for the Royal Society. The brothers produced this etching after a watercolour by Thomas depicting the experience.

And again in 1612:

> Th'Ayre showes such Meteors, as none can see.
> Not onely what they meane, but what they bee.[18]

A poem written a century later suggests how meteoritic events retained their supernatural connotations, at least among the unquestioning general public. 'Meteors' in the 'Autumn' cycle of James Thomson's *The Seasons* (1730) begins with a harrowing description of a meteor shower, 'aerial spears and steeds of fire' signalling 'blood and battle; cities overturned'. But while the masses are afraid,

> Not so the man of philosophic eye,
> And aspect sage; the waving brightness he
> Curious surveys, inquisitive to know
> The causes, and materials yet unfixed,
> Of this appearance beautiful and new.

The poem was reproduced by Sir Richard Phillips in *A Hundred Wonders of the Modern World* in 1834, a 794-page encyclopaedia of natural and man-made curiosities including everything from meteors and Mont Blanc to Stonehenge and St Paul's Cathedral, redolent of the nineteenth-century know-it-all zeitgeist.[19] The previous year had brought a spectacular showing of the Leonid meteor shower and in 1835, Halley's Comet made a brilliant return. The Encke Comet was also in the news; although less visible than Halley's, it appeared as calculated in both 1833 and 1838 and was expected yet again in 1842. This startling series of celestial spectacles prompted proud elucidations of recent scientific discoveries alongside a rash of doomsday fiction. Stones falling from the sky were only lately accepted as proven fact, and other types of cosmic assault might well be in store.[20]

'The Comet' (1839), a short story by S. Austin Jr, was an account of a comet's return which, as the (fictional) savants accurately predicted, struck Earth causing devastating floods.[21] That same year, Edgar Allan Poe (1809–1849) published 'The Conversations of Eiros and Charmion', about two disembodied spirits who discuss humanity's demise as a result of Earth's collision with a comet. Eiros recalls how it was expected to pass dangerously close to Earth, but several days after its appearance:

> men breathed with greater freedom. It was clear we were already within the influence of the comet, yet we lived. We even felt an unusual elasticity of frame and vivacity of mind.[22]

Following a brief summary of current theories of the composition of comets (gaseous) and the atmosphere (potentially flammable), Eiros describes the first signs that something was amiss:

> [People suffered] a rigorous constriction of the breast and lungs, and an insufferable dryness of the skin ... The result of investigations sent an electric thrill of the intensest terror through the universal heart of man.[23]

The comet, they realized, was igniting the atmosphere:

> the whole incumbent mass of ether in which we existed, burst at once into a species of intense flame, for whose

surpassing brilliancy and all-fervid heat even the angels in the high Heaven of pure knowledge have no name. Thus ended all.[24]

In 1843 a non-periodic comet (that is, one that approaches the sun only once or at great intervals) appeared unannounced, trailing the longest, brightest tail yet witnessed. Poe's story received a timely re-release in the *Philadelphia Saturday Museum* (1 April 1843) now entitled 'The Destruction of the World'. In a preface intended to be reassuring, Poe stressed that his story was purely fictional; the great comet had receded and posed no threat, but 'it came unheralded' and who knew what might come next? Faith in God, Poe concluded, 'is by no means inconsistent with a due sense of the manifold and multiform perils by which we are so fearfully environed'.[25] Poe's conflagration would be variously iterated in apocalyptic fiction, including *Cities of the Red Night* (1981) by Poe's admirer William S. Burroughs, where a meteorite falls and 'the whole northern sky [is] lit up red at night, like the reflection from a vast furnace'.[26] But perhaps the earliest contribution to the genre was made by the unknown Egyptian author of 'The Shipwrecked Sailor' (c. 2000 BC), which recounts a series of fantastic adventures, including surviving something resembling a cosmic impact:

Then a star fell.
And because of it these went up in fire.
It happened utterly.[27]

In 1860 yet another non-periodic comet blazed across north America; Walt Whitman's (1819–1892) poem 'Year of Meteors' mentions it while describing a similarly rare meteoritic event that also occurred that year:

. . . the strange huge meteor-procession dazzling and clear
    shooting over our heads,
(A moment, a moment long it sail'd its balls of unearthly
    light over our heads,
Then departed, dropt in the night, and was gone . . .)

American landscape artist Frederic Edwin Church (1826–1900) witnessed the procession over the Catskill Mountains on the night of 20 July 1860 and captured it in luminous oils.[28] Church belonged to the Hudson River School of landscape art, a movement whose reverence for nature was shared with Henry David Thoreau and Ralph Waldo Emerson, contemporaries whom Whitman admired. 'Year of Meteors', written on the eve of America's Civil War, sounded a note of transcendence:

Frederic Edwin Church, *The Meteor of 1860*, 1860, oil on canvas.

> Year of comets and meteors transient and strange – lo! even
> here one equally transient and strange!
> As I flit through you hastily, soon to fall and be gone, what
> is this chant,
> What am I myself but one of your meteors?

Awed by his times, Whitman evoked the *Great Eastern*, a huge iron sail and steam ship (211 metres, 4,000 passengers) that entered New York Harbor in the same elegiac terms as the meteor procession ('Nor forget I to sing of [its] wonder'). As the Industrial Age marched on, accompanied by war and colonial conquest, man-made marvels would soon eclipse celestial ones.

H. G. Wells (1866–1946) is known for his science fiction writing, but this prolific English author wrote numerous non-fiction books and articles on history, politics and science. Extrapolating on current technological and social trends to envisage a near future, Wells predicted sophisticated weaponry such as tanks and the atom bomb, and foresaw the decimation of cities owing to the use of automobiles. His inventive fiction explored themes of invisibility, time and space travel, genetic engineering and most famously, alien invasion, in *The War of the Worlds* (1898). In chapter Two ('The Falling Star') unwitting Earthlings mistake a Martian spaceship for a meteorite. Wells's narrator tells how an astronomer named Ogilvy observed the shooting star and goes hunting for the meteorite he believed 'lay somewhere on the common between Horsell, Ottershaw and Woking'.

> Find it he did, soon after dawn, and not far from the sand pits. An enormous hole had been made by the impact of the projectile, and the sand and gravel had been flung violently in every direction over the heath, forming heaps visible a mile and a half away. The heather was on fire eastward, and a thin blue smoke rose against the dawn.
>
> The Thing itself lay almost entirely buried in sand, amidst the scattered splinters of a fir tree it had shivered to fragments in its descent ... It had a diameter of about thirty yards. He approached the mass, surprised at the size and more so at the shape, since most meteorites are rounded more or less completely. It was, however, still so hot from its flight through the air as to forbid his near approach. A stirring noise within its cylinder he ascribed to the unequal cooling of its surface; for at that time it had not occurred to him that it might be hollow ... suddenly he noticed with a start that ... the ashy incrustation that covered the meteorite, was ... dropping off in flakes and raining down upon the sand. A large piece suddenly came off and fell with a sharp noise that brought his heart into his mouth.[29]

The WAR of the WORLDS
By H. G. Wells
'Author of "Under the Knife," "The Time Machine," etc.

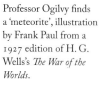

Professor Ogilvy finds a 'meteorite', illustration by Frank Paul from a 1927 edition of H. G. Wells's *The War of the Worlds*.

Ogilvy realizes the cylinder is opening, or trying to open:

> 'Good heavens!' said Ogilvy. 'There's a man in it – men in it! Half roasted to death! Trying to escape!'

Ogilvy wants to help unscrew the cylinder:

> But luckily the dull radiation arrested him before he could burn his hands on the still-glowing metal. At that he stood

irresolute for a moment, then turned, scrambled out of the pit, and set off running wildly into Woking.

Ogilvy is beside himself until he sees 'Henderson, the London journalist, in his garden':

'Henderson,' he called, 'you saw that shooting star last night?'
'Well?' said Henderson.
'It's out on Horsell Common now.'
'Good Lord!' said Henderson. 'Fallen meteorite! That's good.'[30]

Wells was an outspoken socialist and *The War of the Worlds* may be read as his admonishment of a self-satisfied civilization whose pride cometh before its (near) fall:

With infinite complacency men went to and fro over this globe about their little affairs, serene in their assurance of their empire over matter. No one gave a thought to the older worlds of space as sources of human danger . . . [31]

Wells's impressionistic short story 'The Star' (first published in the Christmas 1897 issue of London's *The Graphic*) describes the devastation wrought on Earth when a 'vast mass of matter' emerges from deep space and collides with Neptune, creating a fused incandescent 'star' that grows 'brighter, larger, *nearer*'.

The more playful works of Wells's French counterpart Jules Verne (1828–1905) include *The Chase of the Golden Meteor*, published posthumously by Verne's son in 1908. Many of the 60 novels comprising Verne's *Voyages extraordinaires* adventure series involve space travel, including *Hector Servadac* (1877, published in English as *Off on a Comet*) about a captain of the French Algerian army, who crosses the solar system aboard a comet. As in Wells's works, astronomers are frequent protagonists; *The Chase of the Golden Meteor* features two, vying over who will be known as discoverer of a hurtling mass of solid gold about to land on Earth's doorstep. Verne pokes fun at the presumptive scientific types who argue over who has the right to mine the asteroid 'like two sportsmen shooting over the same preserves'.[32] The men of science are quick to consult the men of finance, speculating as to the asteroid's impact on the market price of gold. When a date and location are calculated for the asteroid's fall, 'innumerable were the people who went mad . . . in a few days the lunatic asylums were found to be too small for the patients they had to accommodate'.

Verne's heroes expound at length on meteoritics, citing the fall at Ensisheim in 1492, predicting the asteroid's trajectories and mineral composition.[33] Scientific detail, accurate or otherwise, lent Verne's narratives authority while serving as a selling point to a reading public amazed by advances in knowledge and technology that must have seemed in the nineteenth century to know no bounds. Staying abreast of findings in astronomy and other scientific developments would become a measure of intellect for fictional characters and readers alike.[34] In Sir Arthur Conan Doyle's *The Valley of Fear* (1914) Sherlock Holmes lauds his arch-rival Professor Moriarty as:

LE GRAND TÉLESCOPE DE L'OBSERVATOIRE DE PARIS

The Paris Observatory telescope, engraving from Amédée Guillemin's *Le Ciel* (1877).

the celebrated author of *The Dynamics of an Asteroid*, a book which ascends to such rarefied heights of pure mathematics that it is said in the scientific press that there is no man capable of criticizing it.[35]

Verne was the first to tie his plots to astronomical discoveries and the technological knowhow of his protagonists, a formula that remains a bestseller. In a Verne-like vein, Dan Brown's *Deception Point* (2001) revived the real-life 1996 Martian meteorite sensation, featuring a fictional meteorite found trapped in

'Night and day
groups were to be
seen [awaiting the
golden meteorite].'
George Roux,
illustration for Jules
Verne's *The Chase
of the Golden Meteor*
(1908).

Arctic ice by NASA, who claim it contains traces of extraterrestrial fossil life. The meteorite becomes a pawn in a Washington
conspiracy to undermine NASA and promote a presidential candidate who will privatize space exploration and exploitation.
Brown's novel is peppered with meteoritic titbits and palaeontologists describe the taxonomy of the alien insects trapped in
the meteorite as identical to a primitive species of terrestrial

'At a distance of 400 yards [observers watch the meteorite fall].' George Roux, illustration for Jules Verne's *The Chase of the Golden Meteor* (1908).

louse. In his author's note, Brown affirms that all the technologies in his novel exist, including gnat-sized flying microbots that send images and sound back to headquarters and charge themselves by landing near a computer screen. Verne, who foresaw the invention of submarines, helicopters and electric batteries, similarly far-fetched gadgets in their time, outlined his writerly aspirations in a journal entry at the age of 28:

Not mere poetry, but analytical fantasy. Something monomaniacal. Things playing a more important part than people, love giving way to deduction and other sources of ideas, style, subject interest. The basis of the novel transferred from the heart to the head.[36]

With these words, the ambitious young Verne seems once again to have accurately predicted the future.

A now familiar plot driver in science (often pulp) fiction, where the action turns on their fall and its real or imagined consequences, meteorites have lent themselves to drama, danger and doom, all wrapped in the redemptive aura of space-age technology. Contemplative, real-life literary references to meteorites are rare, and an anecdote from Antoine de Saint-Exupéry's luminous memoir, *Terres des hommes*, is unique. Pilot for the French Post in the 1920s and '30s, Saint-Exupéry (1900–1944) flew a flimsy wood and canvas plane across the Mediterranean and over the Sahara, where 'a minor accident' forced him to land on 'one of those table lands of the Sahara . . . an isolated vestige of a plateau that had crumbled round the edges'. Saint-Exupéry describes an otherworldly landscape, punctuated by high 'truncated cones', massive calcium deposits from an antediluvian sea, whose surface, he observes, is composed entirely of bleached, 'minute and distinct shells'.

Without question I was the first human being ever to wander over this . . . this iceberg: its sides were remarkably steep, no Arab could have climbed them, and no European had as yet ventured into this wild region.

. . . I lingered there, startled by this silence that never had been broken. The first star began to shine, and I said to myself that this pure surface had lain here thousands of years in sight only of the stars.

But suddenly my musings on this white sheet and these shining stars were endowed with a singular significance. I had kicked against a hard, black stone, the size of a man's fist, a sort of moulded rock of lava incredibly present on the surface of a bed of shells a thousand feet deep. A sheet spread beneath an apple tree can receive only apples; a sheet spread beneath the stars can receive only stardust. Never had a stone fallen from the skies made known its origin so unmistakably.

Saint-Exupéry soon finds a second and third meteorite:

And here is where my adventure became magical, for in a striking foreshortening of time that embraced thousands of years, I had become the witness of this miserly rain from the stars. The marvel of marvels was that there on the rounded back of the planet, between this magnetic sheet and those stars, a human consciousness was present in which as in a mirror that rain could be reflected.[37]

Saint-Exupéry's Saharan experience is reconfigured in *The Little Prince* (1943), whose protagonist falls from an asteroid to the desert where he meets the narrator, a stranded pilot repairing his plane.

'The Meteorite' (1946), a poem by C. S. Lewis, recalls Shelley's 'Ozymandias' (1818):

Two vast and trunkless legs of stone
Stand in the desert. Near them, on the sand
Half sunk, a shattered visage lies . . .
Nothing beside remains.

Lewis's poem describes a mighty meteorite moldering on a hill-top, covered with moss and overgrowth, and remarks how effortlessly Earth 'digests' a 'cinder of sidereal fire', transforming its 'translunary guest' into 'The native of an English shire.' Just as Shelley's broken monolith evoked the vanity of power, so Lewis's poem seems to betray a post-war weariness with destructive marvels while coincidently describing the process of terrestrialization, whereby meteorites erode and are assimilated by Earth's chemistries.

Artist and amateur astronomer Gustav Hahn, son of meteorite collector Otto Hahn, depicted the meteor procession of 9 February 1913, as observed near High Park in Toronto.

With the cross-fertilization of science and literature, meteorite fact and fiction have been conflated, but in at least one instance reality has outstripped both scientific explanations and literary invention. The Tunguska event of 1908, the largest cosmic impact in recorded history, was the earth-shaking explosion of what appeared to be a massive fireball over central Siberia. At 7:17 a.m. on 30 June, a meteorological observatory 860 km away registered the seismic disturbance created by the explosion, which was heard over an area of 1.25 million sq. km. Blast particles rose high above the atmosphere, illuminating the night skies with reflected sunlight for several days. Political turmoil delayed investigations of the remote location (1,000 km north of Irkutsk and Lake Baikal) until 1921, when Russian mineralogist Leonid Kulik (1883–1942) conducted the first fact-finding mission. It was too late in the year to penetrate the snowbound taiga but the eyewitness reports that Kulik collected, of the 'tongue of flame' that 'cut the sky in two', inflamed his imagination.[38]

Local resident S. B. Semeonov, who was in Vanavara, Siberia, around 68 km from the epicentre of the explosion, described the scene:

I had just raised my axe to hoop a cask, when suddenly in the north above Tunguska River, the sky was split in two; high above the forest the whole northern sky appeared to be covered with fire. At that moment I felt great heat, as if my shirt had caught fire ... I wanted to pull [it] off and throw it away, but at that moment there was a bang in the sky, and a mighty crash was heard. I was thrown three sagenes [approx. 6 m] away from the porch and for a moment I lost consciousness. My wife ran out and carried me to the hut. The crash was followed by noise like stones falling from the sky, or guns firing. The Earth trembled, and when I lay on the ground I covered my head because I was afraid some might hit it. At the moment when the sky opened, a hot wind, as from a cannon, blew past the huts from the north.[39]

The blast levelled 2,000 sq. km of virgin woodland in a heartbeat, snapping millions of trees in two, perhaps creating the gunshot sounds Semenov and others heard. No human casualties were reported in the scantily inhabited area, but the blast of intense heat and subsequent fires killed herds of domesticated reindeer belonging to the indigenous Evenki people, who blamed the occurrence on their thunder god, Ogdy. An Evenki family camped north of Vanavara was asleep in their tent when it was 'blown up into the air, together with the occupants. When they fell back to earth ... they saw the forest blazing around them'.[40] Whatever exploded above the taiga on 30 June, whether an icy comet or stony asteroid, as the conflicting theories currently hold, it left scarcely a clue of its origin. Like Daniel Moreau Barringer, who was drilling at Meteor Crater in Arizona at the time, Leonid Kulik would be driven by an obsessive belief in the existence of a meteorite he was destined to never find.[41]

The Tunguska event and Kulik's Siberian odyssey in the path of its destruction are re-imagined in Vladimir Sorokin's *Ice* (2002), the first novel of his 'Ice Trilogy'. Sorokin's narrator, Sasha, was born to a wealthy St Petersburg sugar producer on his family's northern estate the same day as the Tunguska event. His mother blamed his premature birth on the thunder she had both

A caravan of 50 carts transports equipment and provisions through the taiga, March 1929.

heard that morning and 'felt through her fetus, that is, through me'. 'The sky lit up in your honor', his mother told him, referring to the preternatural glow seen throughout the night-time northern hemisphere. As a boy, the precocious narrator develops an interest in astronomy: 'not exactly astronomy itself but the heavenly bodies, hanging in space. Imagining the Universe it was as though I lost myself. And my heart would begin to throb.'[42]

In college Sasha studies physics and befriends a classmate, Masha, who is 'enthralled by a fashionable science – meteoritics . . . Her slanted eyes shining, she spoke enthusiastically about meteorite showers, zodiacal light, iron meteorites with Widmanstätten patterns.' Masha, who planned to accompany Kulik to search for the Tunguska meteorite, introduces him to Sasha, the coincidence of whose birthday wins him an invitation on the next expedition. 'I felt that I *was setting out*', Sasha says.[43]

Kulik's real-life quest began in February 1927 on the Trans-Siberian Railway and then by foot, with horse-drawn sleds carrying provisions towards the fall site. Reaching the trade outpost of Vanavara took more than six weeks of hacking through dense brush, the ground underfoot turning boggy with the spring thaw. Having established a base camp, there were still nearly 60 km to cover, through snow too deep for horses. Enlisting Evenki guides and the reindeer they milked and used for transport, Kulik's party continued its arduous trek on 8 April.

Five days later, they reached the perimeter of the impact zone, where the scorched trunks of trees, some of which had stood 30 m high, were splayed on the ground like burnt matchsticks, as far as the eye could see.[44] Kulik observed that the trees had fallen in a radial pattern away from a swampy basin he called 'the cauldron', which he believed to hold the meteorite. In April 1928 he returned to examine the basin and surrounding pits he thought were craters of smaller meteorites, but floodwaters impaired travel and prevented magnetic readings. The following year Kulik brought drilling tools and stayed throughout a gruelling winter, boring into the basin to no avail. The discouraged Kulik was unable to return to the taiga until 1938, his last trip. He took aerial photographs illustrating the extent of the impact's damage that appeared to confirm the swamp as the fall site, but no traces of meteorite were found.[45]

The 'dead zone', trees felled by the Tunguska impact of 30 June 1908.

In Sorokin's *Ice*, the fictional Kulik recalls his party's 'genuine terror and joy' arriving at the 'dead zone', the perimeter of felled trees: 'Such phenomenal destruction of the taiga could only result from the volition of an enormous meteorite!' he exclaims.[46] Sasha accompanies Kulik's next journey north, travelling through woodlands that remind him 'of the manes of sleeping monsters'.[47] Along the way, he has a dream: 'a finger passed through my ribs and touched my heart' as if to prod it awake. Sasha's 'awakening' begins as he enters the dead zone:

> All the trees lay with their crowns pointed towards me, and their roots towards the setting sun . . . The dead forest impressed one with the scale and force of its sudden demise . . . My heart fluttered, my eyes grew dim. And suddenly I felt wonderful, *terribly* wonderful . . . My eyes filled with tears. An involuntary stream of urine flowed with warmth and tenderness down my legs.[48]

Trench dug around a depression that Leonid Kulik believed to be a crater, *c.* 1929–30.

Sasha grows indifferent to his companions, loses his appetite; his past becomes 'a frozen picture under glass, like an herbarium in a museum'.[49] He separates from the group and discovers a huge mass of ice submerged in the bog, the remains of a comet bearing a message that alters his destiny.

Reborn as 'Bro', Sasha learns that he contains a spark of light shared with 22,999 others; the rest of mankind is asleep, little more than 'meat machines'. His mission is to assemble the 'Brotherhood of the Light', whose combined forces will annihilate Earth and with it all of failed humanity. The Tunguska ice is the vehicle of transformation; those selected for admission to the Brotherhood (all blond-haired and blue-eyed) must survive being bashed in the sternum with a ammer made of it. The delusion of the cult is portrayed as perfect clarity in Sorokin's *Trilogy* (which has been called 'a precise parable for each new stage of Russia's history'); Bro's messianic intensity is seductive yet tinged with the odious.[50] He encounters Kulik ('the meat machine who had led me to the Ice') one last time while visiting a concentration camp: 'It had been captured and was dying, without ever finding out *what* had fallen from the sky to Earth in 1908.'[51] When the Second World War broke out, Leonid Kulik enlisted, was wounded in the Battle of Moscow and was taken prisoner. He died in a concentration camp in 1942, never knowing his work would live on in both scientific debate and literary fiction.

The destructive reality of Second World War weaponry was reflected in 'The Explosion' (1946), a short story by Soviet author Alexander Kazantsev that attributed the Tunguska devastation to the malfunctioning atomic reactor of an alien spaceship. Kazantsev's fiction seeped into popular confabulations regarding the possible cause for the Tunguska event, as did the electricity experiments of Nikola Tesla. Another theory held that an anti-rock composed of antimatter had exploded in the atmosphere; yet another that a black hole pierced the Earth and exited through the Atlantic Ocean. The most plausible causes of the Tunguska event are either an asteroid or a comet, though no one is sure which. NASA sides with the asteroid, as do many meteorite scientists, but the comet theory is also valid. The Russian scientific community nonetheless dismisses as pseudo-science the claims of Vladimir Alexeev of the Troitsk Innovation and Nuclear Research Institute (TRINITY), who in 2010 purportedly located a mass of cometary ice, *à la* Sorokin, beneath the Siberian permafrost with the help of ground-penetrating radar.[52]

Thomas Pynchon sports with the Tunguska event in *Against the Day* (2006), whose title is drawn from the Bible (2 Peter 3:7): 'the heavens and earth [are] . . . reserved unto fire against the day of judgment and perdition of ungodly men'. Pynchon's sprawling narrative relates the adventures of the 'Chums of Chance' who fly a 'hydrogen skyship' (hot-air balloon) called the *Inconvenience*. An army of characters with names like Darby Suckling, Chick Counterfly and Professor Vanderjuice parade through the novel's 1,085 pages, along with Pugnax, the Chums' dog, who offers Pynchon an opportunity to comment on the credulity that once greeted that which falls from the sky:

> like the rest of the [balloon] crew [Pugnax] responds to 'calls of nature' by proceeding to the downwind side of the gondola, resulting in surprises among the surface populations below . . . these lavatorial assaults . . . entered the realm of folklore, superstition, or perhaps, if one does not mind stretching the definition, the religious.[53]

From their lofty perch the Chums contemplate Earth, alighting on philosophical questions in a manner not unlike the epistemological journey of the heroes of Flaubert's satirical *Bouvard et Pécuchet*, who course through field after field of knowledge without harvesting very much. Encountering Spinozism, Bouvard and Pécuchet:

> felt they were in a balloon at night, in icy cold, being swept away on a never-ending course, towards a bottomless abyss, and with nothing around them but the ungraspable, the motionless, the eternal. It was too much. They gave in.[54]

overleaf: (left and right) Iron flecks in the Gold Basin meteorite (ordinary chondrite), found in 1995 in Mohave County, Arizona, photographed in incident light at a magnification of 700×.

Against the Day produces a similar sensation. After 778 pages of blinding erudition, 'a heavenwide blast of light' announces the arrival of the Tunguska plotline. Two characters, Kit and Prance, are in the taiga, when suddenly

> everything, faces, sky, trees, the distant turn of river, went red. Sound itself, the wind, what wind there was, all gone red as a living heart. Before they could regain their voices, as the color faded to a blood orange, the explosions arrived, the voice of a world announcing it would never go back to what it had been … Kit understood for a moment that forms of life were a connected set … He had entered a state of total attention.[55]

The effects of the event were wide-ranging:

> Reindeer discovered again their ancient power of flight, which had lapsed over the centuries since humans began invading the North. Some were stimulated by the accompanying radiation into an epidermal luminescence at the red end of the spectrum, particularly around the nasal area. Mosquitoes lost their taste for blood, acquiring one instead for vodka, and were observed congregating in large swarms at local taverns. Clocks and watches ran backwards … Siberian wolves walked into churches in the middle of services, quoted passages from the Scriptures in fluent Old Slavonic, and walked peaceably out again. They were reported to be especially fond of Matthew 7:15, 'Beware of false prophets, which come to you in sheep's clothing, but inwardly they are ravening wolves.'[56]

But things soon go back to normal, 'the Event receded in memory, arguments arose as to whether this or that had even happened at all'.[57] And the *Inconvenience* flies blithely on for another 300-odd pages.

The Siberian wilderness lends itself to tales of anomalous adventure, a perfect backdrop for the eerie cosmic moments it seems coincidentally to attract. The Tunguska event was followed

The sole depiction of the Sikhote-Alin fireball: oil painting by the Russian artist P. J. Medvedev, who had coincidentally set up his easel near the Siberian village of Iman on 12 February 1947, minutes before the fireball appeared.

by another major impact over the virtually unpopulated Sikhote-Alin mountains.[58] The fireball that exploded at 10:38 am on 12 February 1947 was so bright that it cast moving shadows of the objects on the ground as it coursed across the sunlit sky. The fragmented mass pounded the woodlands with iron meteorites ranging in size from shards to a hunk weighing nearly 2 tons. The most copious fall in recorded history, Sikhote-Alin left 122 craters but caused no injuries.[59] The smaller but significant meteorite Chelyabinsk fall of February 2013 struck one of Siberia's better-populated areas. Among the 1,200 people wounded by the explosion fall-out, many reported temporary blindness from the fireball that at its zenith appeared 30 times brighter than the sun.[60]

Meteorite truth is stranger than fiction, which has only spurred authors to spin taller tales. But what of meteorite art? Although the literary use of meteorites is of greater historic breadth than that of the visual arts, contemporary artists are contributing to what may emerge as a new genre. Maurizio Cattelan's *The Ninth Hour* (1999) is a lifelike sculpture of Pope John Paul II lying dead, having been hit by a meteorite. Meteorites in fact put paid to religious notions regarding Earth and the solar system but Cattelan was less interested in allegorizing the early history of meteoritics than in 'the collision of religion and blasphemy'. Not everyone appreciated his intentions. At a Warsaw showing, two Polish nationalists pushed the meteorite off the pope and tried to stand him upright, a response Cattelan lauded as 'a sort

Sikhote-Alin meteorite, nearly split by the tensions of flight through Earth's atmosphere; 40 × 35 × 24 cm, 99.5 kg.

Maurizio Cattelan, *The Ninth Hour*, 1999, mixed media.

overleaf:
Thin section of the NWA 969 LL7 meteorite photographed in cross-polarized light with the addition of a 1/4 wave retardation filter at a magnification of 160×. This extremely rare chondrite shows fluid trains trapped in small round cavities in olivine grains . . . 4.5 billion-year-old bottled water.

of upside down miracle: salvation wasn't coming from the sky but from the earth, from the people'.[61]

Perhaps the first artist to use meteorites as medium, Cornelia Parker created *Shirt Burnt by a Meteorite* (1996) with a piece of Gibeon iron (from Namibia) she heated with a blowtorch. For *A Meteorite Lands on Buckingham Palace* (1998) Parker warmed a meteorite on her kitchen stove in order to scorch locations on a London map. 'An alien object from space, the meteorite embodies the fear of the unknown, fear of the future', she explained; 'in this sense this is an apocalyptic work for the end of the millennium'.[62] Parker also tossed a lunar meteorite

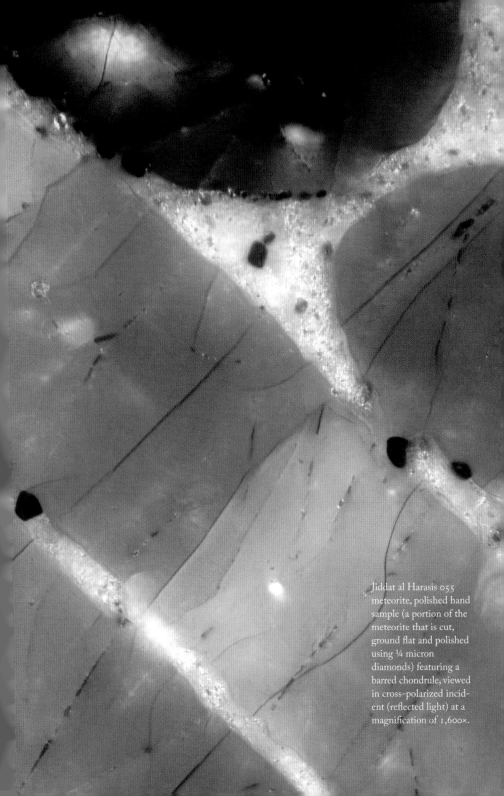

Jiddat al Harasis 055 meteorite, polished hand sample (a portion of the meteorite that is cut, ground flat and polished using ¼ micron diamonds) featuring a barred chondrule, viewed in cross-polarized incident (reflected light) at a magnification of 1,600×.

into a lake and commemorated the act with a sign reading: *At the Bottom of This Lake Lies a Piece of the Moon* (2000). On another occasion, she set off fireworks containing pulverized meteorite from the roof of the Rotunda Building in Birmingham, the closest Parker has so far come to her ambition of sending a meteorite back into space.

Inspired by the meteorite fall in Sylacauga, Alabama, in 1954, German photographer Regine Petersen began *Find a Falling Star* in 2009, a series of works comprised of archival and original images. 'As a worldless and indifferent object,' she writes, 'the meteorite also serves as a canvas for our projective desires, fantasies and fears.'[63] For the 2012 London Olympics 'Road Show' Katie Paterson exhibited a 110-kg piece of the Campo de Cielo iron (Argentina) that she had melted down and recast. 'The artist domesticates the cosmos's immensity', remarked one critic, 'she gives the unfathomable a human scale, putting it within our reach.'[64] In 2010, critic and conceptual artist Jonathon Keats grew cacti and potatoes in mulch containing a powdered chondrite and later sold water distilled with Martian and lunar meteorites for U.S.$30 per bottle as part of his 'Local Air and Space Administration' (LASA) project. Keats, who has copyrighted his mind as a sculpture, says, 'My system is safer than any other means of space exploration . . . you just

Jonathon Keats, *Astronauts (Local Air and Space Administration)*, 2010, potatoes raised in water mineralized with Martian shergottite NWA 1195. 'On the nature of Mars, potatoes now know more than we do.'

sit back and sip.'[65] In more ways than one, as Chinese performance artist Li Wei has said, 'art is to play with gravity'. In 2003 he photographed himself crashing into a car windshield in an acrobatic *mise en scène* (see p. 160). 'I act like a meteorite falling into something [to] indicate a state of danger, just like the condition of human beings.'

Most of us will never see a fireball explode or a meteorite fall or, hopefully, be incinerated by a cosmic impact; our mental image of such events is shaped from a pastiche of scientific data, reportage and creative renderings. *Sites of Impact* (2009), a book

Jonathon Keats,
Spacecraft (Local Air and Space Administration),
2010, chair set on slices of the Campo de Cielo meteorite and slippers clad with fragments of the Nantan meteorite.

of photographs of Earth's craters by Stan Gaz, restores clarity and scale to the contemplation of these realities. 'Poised between engagement and detachment', Gaz's aerial images offer a God's-eye view of topographics sculpted by forces that tempt the intellect and defy the imagination. In sombre, solarized black-and-white prints, the craters are presented chronologically, beginning with one formed 900 million years ago and ending with another that is just 5,000 years old. Taken together, Gaz's 'footprints of the stars' present us with Earth as time's canvas.[66] Science may measure or explain these cosmic processes without ever approaching the grander, interactive reality of our planet's life in space, whose meaning, in the end, is no more or less than that which we assign it. And that is perhaps where meteorites enter the realm of philosophy, illumined by literature and art.

Stan Gaz, *Henbury Crater Field, Northern Territory, Australia*, 2009, solarized black-and-white print.

5 Strange Landings

Egypt's surprisingly modest collection of geological treasures is housed in a careworn hangar in Cairo by the Nile, its entrance flanked by unidentified boulders. The drop-ceilinged interior, lit with sputtering fluorescent tubes, holds rows of display cabinets from the exhibit's earlier home in a beautiful building from 1908 that was torn down in the 1980s and replaced with a subway station. The fossil collection is unexciting, except for the grand metre-high pelvic girdle of an *Arsinoitherium*, a kind of twin-horned rhinoceros that roamed a tropical Egypt 35 million years ago. The minerals and rocks are displayed like miniature strewn fields, tossed unceremoniously into the case with laminated labels propped beside them. There are piles of glossy black tektites, natural glass formed from terrestrial debris fused by the extreme heat of a cosmic impact. Among them are samples of a chartreuse tektite called 'Libyan Desert glass', a scarab-shaped lump of which was found in the tomb of the boy-king Tutankhamen (*r. c.* 1332–1323 BC), as the centrepiece of a splendid gold pendant.[1]

In a dim corner of the hangar stands a wooden box containing a moon rock hand-delivered to Earth by NASA astronauts and donated to Egypt by former United States president Richard Nixon. Beside it was what I came to see, a famous meteorite from a shower that pelted Nakhla, a farming village near Alexandria, in June 1911. Over half a metre long and roughly fish-shaped, the black Nakhla sat in an ordinary office bookcase with a glass front, its pitted surface filled with sandy dust. A

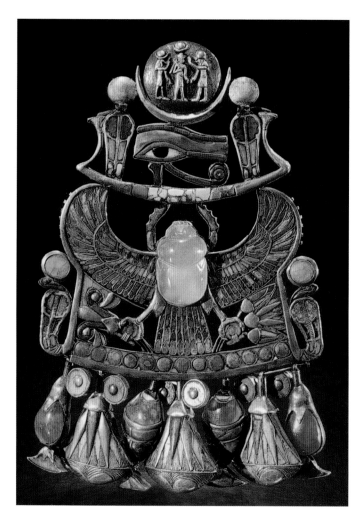

Gold pendant with tektite (impactite glass) centrepiece, found in the tomb of Tutankhamen.

specimen of the Gebel Kamil meteorite resembling a crumbling Christmas pudding shared the shelf. The Nakhla is a piece of Mars, ejected into space by a collision on that planet's surface some 10 million years ago, a fact not mentioned in the display. Observing this alien rock in the flickering gloom, lying on a sheet of blue plastic such as that used to line kitchen cabinets, one could not help thinking this was a mighty odd way for it to end up. I had yet to learn of the bizarre fates reserved for some of the meteorites whose odyssey brought them to Earth.

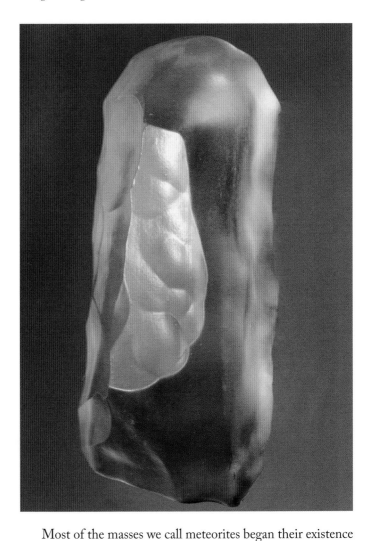

Most of the masses we call meteorites began their existence as the leftover rubble from the solar system's coalescence or else the detritus produced by interplanetary collisions. For millions or billions of years, an infinity in human terms, these relics reel through space, until the laws of motion send some our way. Many go unnoticed; blending with terrestrial rocks, they disappear. Other meteorites have a more conspicuous destiny: arriving on Earth marks the start of yet another journey, beginning with a name assigned in accordance with the locality nearest the fall.

A specimen of 'Libyan Desert glass' formed when an asteroid impact in the Sahara 26 million years ago ejected molten sand that solidified into glass as it fell to Earth.

Cosmic immigrants to Texas include Happy, Noodle, Tarzan, Winnie, Kickapoo and Farewell.[2] These orphaned fragments of failed planetoids, the ultimate outsiders, have been intimately embraced by us humans not just for what they are but also for what we wished for them to be. Whether as the subjects of scientific scrutiny, as collectibles, objects of art or in a variety of other guises, the use of meteorites throughout history reflects a telling range of needs and propensities, from the pragmatic to the fanciful.

One of the more compelling indications of the insider status awarded to meteorites is that they have been ground and eaten. The belief in their curative powers made European demand for meteorites so great in the late Middle Ages that tricksters carried 'betyl quackery' to its 'reprehensible conclusion' by hawking ordinary stones as cure-alls in the place of the genuine article.[3] It has been suggested that China possesses a relatively small number of meteorites from historic falls because they 'went directly into the pharmacopeia'.[4] In 1886 villagers reportedly ingested a meteorite that fell in Novo Urei, Russia. 'They may have regretted it because [it was] the first known example of the rare variety [of meteorite] called *ureilites* . . . containing microdiamonds, which could have done great damage to their teeth.'[5] In 1992 in Mbale, Uganda, people suffering from HIV/AIDS ate a meteorite that fell near their homes, believing it was a remedy sent from heaven.[6]

Meteorites have been recast as other desirables, including the sort of ornamental armament men love. A meteorite blade accompanied Tutankhamen to his tomb.[7] The blacksmiths

Knife made for Mughal emperor Jahangir, partially of meteoric iron, with gold inlay, 1621, India.

Watch face made of meteorite.

serving Jahangir, the fourth Mughal Emperor of India, forged him a dagger by mixing one part terrestrial iron with three parts of a 'lightning iron' that fell in 1621. Jahangir's weapon, with its intricate gold-inlay handle, was said to cut beautifully.[8] Meteorites have also served the human penchant for adornment; perhaps the oldest known examples are the tubular beads fashioned from meteoritic iron found in a 3300 BC burial in Gerzeh, Egypt. Beads and earrings that were likewise cold-worked (hammered) from meteorite have been found in Hopewell burial sites (Ohio and Illinois, *c.* 336 BC).[9] Nowadays all sorts of meteorite jewellery are available online: meteorite watch faces, cufflinks and wedding rings ranging in price from the hundreds to thousands of dollars. Among the high-end novelty items made from meteorites are guitar picks, something no self-respecting rock star should presumably be without. Winners of the 2014 Sochi Winter Olympics were awarded gold medals inlaid with a slice of meteorite.[10]

While meteorites have often fallen into the hands of privilege, others' fates are more prosaic, like the one a Tennessee couple painted green and used as a door-stop.[11] Farmers who found iron-rich meteorites while ploughing their fields typically took them straight to the blacksmith, where they were reborn as horseshoes, nails and farm tools. In at least one case, a meteorite became the anvil itself.[12] But meteorites have also been the hammer, in a manner of speaking: those that strike man-made objects are dubbed 'hammerstones'. Rather than agents of destruction, hammerstones have the power to transform the mundane things they hit into sought-after 'impactifacts'. The most famous impactifact, exhibited in museums worldwide, is the Peekskill

Peekskill Car, with Ray Meyer, who acquired a portion of the meteorite that struck it on 9 October 1992.

Car, a 1980 red Chevrolet Malibu that was parked beside its owner's home in Peekskill, New York, in October 1992 when a 13-kg meteorite crash-landed on its boot (trunk). The car's eighteen-year-old owner, Michelle Knapp, had recently bought the Chevy for U.S.$300. But the meteorite technically belonged to Michelle's mother, since it fell on her property. Mrs Knapp sold portions of the meteorite to three different collectors and a fourth soon acquired the car for U.S.$10,000.[13] A more evocative impactifact was created on 26 March 2003 in a suburban home in Chicago, Illinois, when a grapefruit-sized meteorite smashed through the ceiling and struck *Civilization*, a video game lying on the floor.[14] The lines of Carl Sandburg come to mind:

Lay me on an anvil, O God
Beat me and hammer me into a crowbar
Let me pry loose old walls
Let me lift and loosen old foundations.[15]

Meteorites have been known to strike unsuspecting humans, including Ann Hodges, who was napping on her couch in

Sylacauga, Alabama, on 30 November 1954 when a 4-kg chondrite perforated her wooden roof, bounced off a radio and slammed into her hip. Asked how she felt about this rude awakening, Mrs Hodges replied: 'Bruised'.[16] A 3-g meteorite from the Mbale fall of 1992 hit a Ugandan boy on the head but, slowed by its passage through a leafy banana tree, caused no harm.[17] In August 2002 in Yorkshire, a meteorite landed on the foot of a teenager who was stepping out of a car. The family planned to keep it as a memento. 'After all,' said the girl's father, 'it's not every day you get hit by a meteorite.'[18] Indeed, the odds are quite slim if the relatively few documented cases are anything to go by.

Regine Petersen, *Ann*, 2012/1954, archival pigment ink print, from the series *Stars Fell on Alabama*.

The Bible tells of fleeing Amorite tribesmen who were killed by 'great stones from heaven', though some versions refer to 'hailstones' (Joshua 10:11). The Chinese astronomical annals contain more precise references to 'iron rain' and 'falling stars', with the earliest mention of meteorite casualties dating to AD 616. When 'a large shooting star like a bushel fell into the rebel Lu-Ming's camp', a 'wall-attacking tower' was destroyed and 'more than ten people' crushed to death, probably beneath the falling tower. Meteorite showers recorded in the 1300s reportedly killed dozens of people and animals, but a reference to the deaths of tens of thousands in Qingyang, Shaanxi province (now in Gansu province) in 1490 seems improbable, especially since the largest meteorites were 'like goose eggs' and the smallest 'like water chestnuts'. A meteorite that fell in a marketplace in Ch'ang-shou County in 1639 destroyed 'several tens' of houses, killing as many people. Houses were crushed in 1874 and 1907, along with some of the occupants. In 1915 a meteorite struck a woman's shoulder and reportedly 'tore off her arm'. Assessments of the Chinese records and other related data has led researchers to suggest that someone somewhere will be hit by a small (100-g) meteorite only once every fourteen years.[19]

Earth's current population of approximately 7 billion in fact occupies but a small fraction of our planet's surface, if each inhabitant is allotted a square metre. By some calculations you are more likely to win the lottery twice, or else flip a coin that comes up heads 44 times, than to be hit by a meteorite.[20] This has not stopped some insurance companies from balancing the risks with the eventual costs of impact-related damage. Anyone wishing to determine the odds of their house being subjected to this particular form of *force majeure* may consult a tabulator on a real estate blog mounted shortly after the Chelyabinsk fall of 2013.[21] Based

Stamp commemorating the meteorite fall near Jilin City of 8 March 1976, issued by the People's Republic of China on 21 June 2003. Meteorites have appeared on the postage stamps of at least 27 countries.

solely on size, a home of 300 sq. m has a one in 10,981,348,180,117 chance of being struck.

Probabilities are one thing but reality always seems to have the last laugh. In the summer of 2011, a family home on the outskirts of Paris was struck by an egg-sized meteorite. One of only 40 falls recorded in France over the last four centuries, the event was rendered unique by the fact that the family's name was Comette.[22] In December 1984 a meteorite delivered itself to a mailbox in Claxton, Georgia, punching a hole through the top. Perhaps the most serendipitous fall occurred in Madrid on 10 February 1896, when a meteorite landed near the entrance to Madrid's National Museum of Natural Sciences

Carutha Barnard and neighbours examining the meteorite that struck her mailbox in Claxton, Georgia, on 10 December 1984. Acquired by the Macovich Collection, the mailbox sold at auction for U.S.$82,000 in 2007.

and was collected by Augusto Arcimís, director of the museum's Meteorological Institute.[23]

Science has come a long way towards solving two burning questions posed by meteorites, namely where exactly they come from and where they will touch ground. It was Czech astrophysicist Zdeněk Ceplecha (1929–2009) who devised the equipment and methodology for determining both. On 7 April 1959 two photographic stations 40 km apart recorded the flight of a fireball that exploded into several fragments near Příbram, 50 km southwest of Prague. Ceplecha used these images to calculate the fireballs' trajectory back to the orbit of its parent body in the main asteroid belt and to identify the probable fall site. The Příbam meteorites were the first to be 'instrumentally observed' and have their 'full story' documented from space to Earth.[24] Following the Příbram fall, Ceplecha founded the European Fireball Network (EFN), establishing additional photographic stations to broaden the coverage of the night skies.[25]

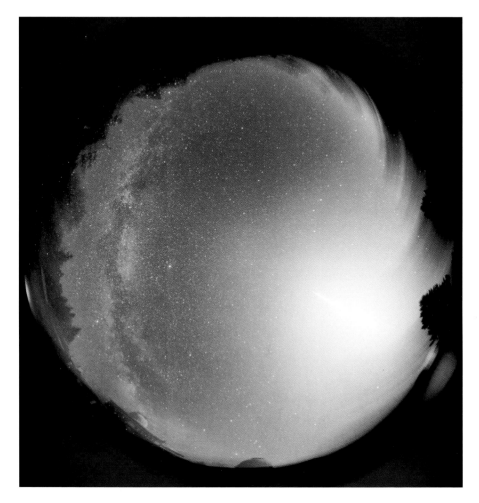

In recent years one of Ceplecha's former students, Pavel Spurný, current head of the Interplanetary Matter department of the Czech Academy of Science's Astronomical Institute, refined his mentor's methods, automating the EFN's 'all sky' cameras and modifying the format from film to high-resolution digital. Spurný operates a photographic station and coordinates the rest of the EFN network from his base the Ondřejov Observatory, built atop a wooded hill around 40 km southeast of Prague. On the morning of 8 May 1991 he checked the results of the previous night's photographic vigil and discovered that an exceptionally brilliant

The Benešov fireball (visible as a streak in the centre of the highly exposed area) was bright enough to clearly illuminate the domed structures of the original 19th-century observatory and surrounding trees. Digital image taken with the 'all-sky' automated camera of the European Fireball Network.

fireball had exploded around midnight local time above Benešov, just 27 km from the observatory, literally within eyeshot. 'I was sleeping when it happened, but I lost a lot of sleep over that fireball afterwards,' says Spurný, who ran the numbers to determine the fall site coordinates and set off with his colleagues to search for the meteorites. For weeks, they combed the ground and trees and found nothing. At that time, only three meteorite falls had been instrumentally observed (at Příbram, Lost City, Oklahoma, and Innisfree, Alberta).[26]

Working with the EFN, Spurný subsequently assisted the establishment of photographic stations throughout central Europe. Situated 100 km apart, they cover an area of about 1 million sq. km and automatically photograph the entire visible sky, their fisheye lenses aimed at the zenith. The night-long exposures obtained from these simultaneous observations form the basis of three-dimensional reconstructions yielding data regarding orbit, trajectory, velocity, luminosity, the moment of explosion and the number, size and composition of fragments produced, in addition to the coordinates of the fall site. Spurný also helped set up photographic stations in the Nullarbor desert of southwest Australia, where the flight of a fireball was recorded on 20 July 2007, and the Bunburra Rockhole meteorites were subsequently recovered in the predicted area with 'X marks the spot' precision.[27] In 2011 Spurný returned to the Benešov fireball twenty years after the event, recalibrating the trajectories to include a wind factor and other equation refinements. The original coordinates, he found, had been 280 m off. This time the team found the meteorites in a field, a bit weathered, but exactly where they were expected.[28]

So far fewer than a dozen meteorites have had their 'full story' told from a scientific point of view. From a historical perspective the most compelling part of the plot is the location and circumstances of the fall, since meteorites not only put otherwise obscure places in the spotlight, but they become part of a narrative linking the locality to a greater story. Such was the case with Berlanguillas, a village near the town of Burgos, Spain, where a meteorite shower occurred on 8 July 1811. Against the

background of the Peninsular Wars, farmers mistaking the exploding fireball for cannon shot and the falling stones for 'bullets whistling past their ears' assumed a battle had broken out with the occupying French troops. A General Dorsenne filed a report to the French Institute and sent a sample of the meteorites to the Museum of Natural History in Paris, and that might have been the end of it. But in 2011 the city of Burgos decided to celebrate the bicentennial anniversary of the meteorite's fall with a widely advertised daylong conference on 'the importance of meteorites for the origin and evolution of Earth and life'.[29]

One of the conference's outcomes was the commissioning of a replica of the meteorite that was sent to France. Factum Arte, a Madrid-based team of artists, conservators and technicians specializing in copies of cultural artefacts, made two versions, one to scale for a Burgos museum and another four times larger

Superbolide, Benešov, Czech Republic, 7 May 1991. Digital image taken with the 'all-sky' automated camera of the European Fireball Network from Telč.

for public display elsewhere. The Berlanguillas was thus the first meteorite to have been painstakingly recorded and recreated (using CNC routers) to preserve an exact copy for posterity. It currently shares this honour with no less than the tomb of Tutankhamen and Paolo Veronese's painting *The Wedding at Cana*, which were also copied by Factum Arte.[30] A lesser attempt at verisimilitude was made by the Japanese chocolatier L'éclat, who designed a boxed set of eight meteorite chocolates in 2013, each inspired by a famous fall and containing fillings with flavours like grapefruit, Earl Grey tea and hazelnut praline.[31] Another

The making of a meteorite: Factum Arte technician creates a fusion crust.

toothsome facsimile was produced following the April 2012 meteorite fall near Sutter's Mill, California, by a local candy manufacturer who fabricated a mounded confection of coconut rolled in black sugar to resemble a fusion crust.[32]

Having conferred wealth and fame on people and places, it is little wonder that meteorites have proved congenial to today's commercial consciousness. Like any famous brand, they not only sell themselves but things associated with them. The Dogfish Head Brewery in Delaware produces 'Celest-Jewel-Ale', reportedly using crushed lunar meteorites 'which help the fermentation process, ironically adding an earthier texture to the drink'.[33] Guerlain's face powder 'Météorites' does not contain space rock, but claims to 'hold the secret to Stardust technology – a light-creating polymer, which transforms light invisible to the naked eye into a pure and endless glow on the skin'.[34] Cross's 'Meteorite Champagne gel ink rollerball pen and pencil set' promises a 'futuristic look and feel' and comes with a 'perpetual life warranty irrespective of age'.[35]

Close-up of the stone meteorite NWA 5717, whose chondrules contain the raw ingredients of our solar system.

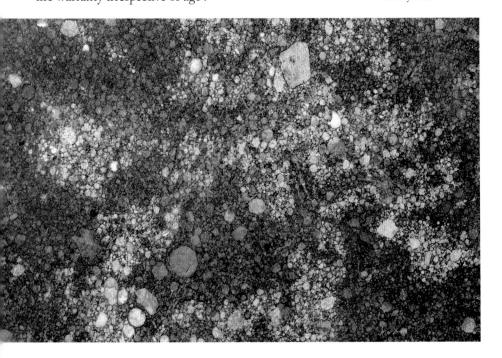

Artist's impression of the closest known planetary system to our own, Epsilon Eridani. Observations from NASA's Spitzer Space Telescope show that the system hosts two asteroid belts.

Considering the ways they have been perceived and put to use, meteorites may yield as much information about us Earthlings as of the origins of Earth itself. And who knows? Maybe pieces of our planet have pierced the skies of other worlds, raised as many questions and given flight to fancy. For if meteorites tell us anything, it is that the universe is grand past all imagining and that they, like us, exist amid its endless possibilities.

REFERENCES

1 Alpha and Omega

1 Report from Captain Edward Topham, landowner of Wold Cottage, Wold Newton, *Gentleman's Magazine*, 1 July 1797, v. 67, part 2, p. 550. Available at www.archive.org.
2 Ibid.
3 Ursula B. Marvin, 'Meteorites in History: An Overview from the Renaissance to the 20th Century', from *The History of Meteoritics and Key Meteorite Collections: Fireballs, Falls and Finds*, ed. G.J.H. McCall, A. J. Bowden and R. J. Howarth, Geological Society, London, Special Publications 256 (London, 2006), p. 31.
4 Ursula B. Marvin, 'Ernst Florens Friedrich Chladni and the Origins of Modern Meteorite Research', *Meteoritics and Planetary Science*, 31 (1996), p. 555.
5 J. G. Burke, *Cosmic Debris: Meteorites in History* (Berkeley, CA, 1986), p. 11.
6 A. Motte, trans., *Newton's Principia: The Mathematical Principles of Natural Philosophy* (New York, 1825), p. 35, available at archive.org.
7 Massimo D'Orazio, 'Meteorite Records in Ancient Greek and Latin Literature: Between History and Myth', *Myth and Geology*, ed. L. Piccardi and W. B. Masse, Geological Society, London, Special Publications 273 (London, 2007), p. 216.
8 Marvin, 'Meteorites in History', p. 32.
9 Ibid., pp. 37–8.
10 Saxe Commins and Robert N. Linson, eds, *The Philosophers of Science* (New York, 1954), p. 10.
11 Marvin, 'Meteorites in History', p. 32.
12 Ibid.
13 Matt Salusbury, 'Meteor Man', *Fortean Times*, CCLXV (August 2010).
14 E. Darwin, 'The Botanic Garden, A Poem, in Two Parts' (1791), available at http://ota.ahds.ac.uk.
15 M. Beech, 'On Meteors and Mushrooms', *Journal of the Royal Astronomy Society*, Canada, LXXXI/2 (1987), pp. 27–9.

16 Ibid.

17 Ibid.

18 The 1847 edition of Pliny's *Natural History* notes: 'for a long time the fall from the sky, of what are denominated Meteorolites, was deemed too preposterous to be believed; but since the facts are no longer doubted, the instances recorded by Pliny become valuable evidences of their antiquity.' Available at www.archive.org.

19 The 'blood rain' that fell on Kerala, India, from July to September 2001 was initially thought to be related to an exploded fireball. A state-commissioned study concluded it was instead coloured by an airborne species of local algae. Kerala has experienced several red rains, including in 2012.

20 Marvin, 'Meteorites in History', p. 50.

21 F. Bacon, *Novum organum*, Book 1, Aphorisms, available at www.archive.org.

22 K. Yau, P. Weissman and D. Yeomans, 'Meteorite Falls in China and Some Related Human Casuality Events', *Meteoritics*, XXIX (1994), pp. 864–71.

23 Ibid.

24 Alistair McBeath and Andrei Dorian Gheorghe, 'Meteor Beliefs Project: Notes from Some Early Medieval Annals', *WGN, the Journal of the International Meteor Organization*, XXXV/6 (2007), p. 141; Colin S. L. Keay, 'Progress in Explaining the Mysterious Sounds Produced by Very Large Meteor Fireballs', *Journal of Scientific Explanation*, VII/4 (1993), p. 338; Ted Nield, *Incoming* (London, 2011), p. 12.

25 Keay, 'Progress in Explaining the Mysterious Sounds', pp. 337–8.

26 Ibid., p. 338.

27 Ibid., p. 340.

28 Ibid. See chapter Three of the present work for a discussion of Nininger's career in meteoritics.

29 The editors of *An Introduction to the Study of Meteorites* (British Museum, Mineral Department, 1894) noted how 'witnesses of [meteorite falls] have been treated with the disrespect usually shown to reporters of the extraordinary, and have been laughed at for their supposed delusions', p. 17.

30 Marvin, 'Meteorites in History', p. 34.

31 Chladni's seminal work, *Die Akustick* (1803), was translated and published in France in 1809 by French emperor and amateur scientist Napoleon Bonaparte.

32 Marvin, 'Ernst Florens Friedrich Chladni', p. 581.

33 Author's conversation with astrophysicist Amr Zant, 24 January 2014.

34 Burke, *Cosmic Debris*, p. 11.

35 Marvin, 'Meteorites in History', p. 47.
36 Heinrich Wilhelm Olbers, who discovered the Pallas and Vesta
    asteroids (in 1802 and 1807 respectively), initially thought meteorites
    came from lunar volcanoes. In 1837 he published his support for
    Chladni's cosmic origin theory, ten years after Chladni's death.
37 Marvin, 'Meteorites in History', p. 50.
38 Matthieu Gounelle, 'The Meteorite Fall at L'Aigle and the Biot
    Report: Exploring the Cradle of Meteoritics', in *The History of
    Meteoritics*, p. 78.
39 Helen Sawyer Hogg, 'Out of Old Books', *Royal Astronomical Society
    of Canada Journal*, LVI/5 (1962), p. 216. Meteor showers occur when
    the Earth intersects swarms of meteoids left from the passage of
    a comet orbiting the sun that vaporize in the upper atmosphere at
    altitudes of 64–128 km.
40 Ibid., p. 220.
41 Ibid., p. 217.
42 The article mocks astronomers' esoteric instructions to their
    acolytes: 'The observer, provided with a piece of chalk, will embrace
    [a] tree with his clasped hands at full arm's length, the head and
    body being held erect. At the appearance of a Meteor the body will
    be swung about until the bole of the tree or post intersects upon the
    heavens, the central point of the Meteor's path, and there, without
    deranging body or eye, he will chalk at the center of the tree's
    face, a small figure . . .', in A. McBeath, G. J. Drobnock and A. D.
    Gheorghe, 'Meteor Beliefs Project: Seven Years and Counting',
    WGN, the Journal of the International Meteor Organization, XXXVIII/2
    (2010), p. 52.
43 Translated to French in 1929, the first (abridged) English edition
    of *The Biosphere* was published by Synergetic Press (1986) with a
    preface by Prof. Evgenii Shepelev, pioneer in closed-systems studies.
    In 1920, Vernadsky and a mineralogist colleague persuaded Lenin
    to establish the world's first nature reserve (in the southern Urals)
    for the purpose of scientific study.
44 Marvin, 'Meteorites in History', p. 65.
45 Analysis of a calcium-aluminum-rich inclusion (CAI) in the
    Northwest Africa 2364 chondritic meteorite reset the age of the
    solar system at 4.5682 billion years, predating previous estimates
    by up to 1.9 million years. CAIs are thought to be the first solids to
    condense from the cooling protoplanetary disk during the birth
    of the solar system. Audrey Bouvier and Meenakshi Wadhwa, 'The
    Age of the Solar System Redefined by the Oldest Pb–Pb Age of
    a Meteoritic Inclusion', *Nature Geoscience*, III (2010), pp. 637–41.
46 Lucretius, *On the Nature of Things*, quoted in *The Philosophers of
    Science*, ed. Robert N. Commins (New York, 1954), p. 14.

47 Matthias Willbold, Tim Elliott and Stephen Moorbath, 'The Tungsten Isotopic Composition of the Earth's Mantle before the Terminal Bombardment', *Nature*, 477 (2011), pp. 195–8.

48 'Where Does All Earth's Gold Come From? Precious Metals the Result of Meteorite Bombardment, Rock Analysis Finds', *Science News Daily*, 9 September 2011.

49 Bill Bryson, *A Short History of Nearly Everything* (London, 2004), p. 238.

50 B. Steigerwald, 'NASA Finds Clues to a Secret of Life in Meteorite Dust', *NASA Goddard Space Flight Center*, March 2009, online. A collision between two asteroids captured by the Hubble Telescope produced a meteorite that fell in Sudan in 2008. In 2010 it was found to contain nineteen amino acids that survived the 1,000°C heat produced by the impact.

51 NASA also lists sixteen 'provisional' moons: http://solarsystem.nasa.gov, accessed 2 April 2015.

52 'Asteroids: Read More', at http://solarsystem.nasa.gov/planets/index.cfm, accessed 2 June 2015.

53 The scale analogy from Bryson, *A Short History of Nearly Everything*, p. 45.

54 Ibid., p. 49. These cosmic odds are reminiscent of the Tibetan notion that the consciousness released at a person's death would have about the same chance of finding a new body as a turtle surfacing in one of several rings floating atop an ocean.

55 Author's conversation with Amr Zant, 24 January 2014.

56 K. Zahnle and N. H. Sleep, 'Impacts and the Early Evolution of Life', in *Comets and the Origin and Evolution of Life, Advances in Astrobiology and Biogeophysics*, ed. Paul J. Thomas et al. (Berlin, 2006), p. 289.

57 Jon Larsen, 'A Hunt for Micrometeorites', *Meteorite: International Quarterly Magazine for Meteorites and Meteorite Sciences*, XVIII/2 (May 2012), pp. 31–6.

58 Nield, *Incoming*, p. 95. A student and then a colleague of Derek Ager, Neild speaks at affectionate length of his mentor's independent thinking and achievements.

59 Ibid., p. 100. The microdiamonds, an example of shock metamorphism, were formed from carbon in the meteorite under the pressures of impact. In Foote's day it was theorized that the meteorite was a fragment of a body large enough to form diamonds in its core.

60 Ibid., p. 102. William D. Boutwell, 'The Mysterious Tomb of a Giant Meteorite' in *National Geographic* magazine, LIII (1928), features an undated photograph of Gilbert's mud-ball experiments, credited to the U.S. Geological Survey, p. 727.

61 Nield, *Incoming*, p. 104.
62 Drew N. Barringer, 'A Grand Obsession: Daniel Moreau Barringer and Meteor Crater', *Meteorite, International Quarterly Magazine for Meteorites and Meteorite Sciences*, xvi/4 (November 2010), pp. 8–9.
63 Ibid., p. 9.
64 Nield, *Incoming*, p. 106.
65 Barringer, 'A Grand Obsession', p. 12.
66 Boutwell, 'The Mysterious Tomb of a Giant Meteorite', p. 722.
67 Barringer, 'A Grand Obsession', pp. 11–12. Aside from two mining-related books that became standard references in mining and engineering schools, Barringer authored two scientific papers supporting Meteor Crater's impact origin, noting the finely pulverized silica found within it (later understood to be formed by shock metamorphosis) and large quantities of magnetic iron oxide (meteorites) around its rim that were inconsistent with known volcanic materials, none of which were found in the vicinity.
68 Recognizable impact craters were rare in Shoemaker's day; as of 2012, 180 had been identified worldwide, mostly by satellite imagery. The oldest (3 billion years) discovered in Greenland in 2012, is nearly 100 km wide. The largest and second oldest is the 300-km-wide Vredefort crater in South Africa.
69 Bryson, *A Short History of Nearly Everything*, p. 240.
70 Nield, *Incoming*, pp. 96–8.
71 Ibid., p. 98.
72 Ibid., p. 117.
73 V. I. Vernadsky, 'Evolution of Species and Living Matter', trans. Meghan Rouillard. Text first appeared as an appendix to the French edition of *The Biosphere* (1929).
74 H. G. Wells, *Outline of History* [1919] (New York, 1956), vol. 1, p. 41. See chapter Four of the present work for a discussion of Wells's literary work.
75 L. W. Alvarez, F. Asaro and H. V. Michel, 'Extraterrestrial Cause for the Cretaceous–Tertiary Extinction', Science, ccviii/4448 (6 June 1980), pp. 1095–108.
76 Byars heard Alan Hildebrand present a paper on a possible Caribbean impact site and informed him of a study of 1981 by Pemex Oil geophysicist Glen Penfield which theorized that the magnetic and gravitational anomalies noted by Pemex scientists indicated an impact crater. When Hildebrand joined forces with Penfield to present the geological evidence, the theory was accepted. See Hildebrand and Penfield et al., 'Chicxulub Crater: A Possible Cretaceous/Tertiary Boundary Impact Crater on the Yucatán Peninsula, Mexico', *Geology*, xix/9 (September 1991), pp. 867–71. The dimensions given are 'transient', lasting briefly after impact

before the crater's sides collapsed. Lying mostly beneath the seabed with a portion extending inland, Chicxulub's 'final' dimensions are approximately 154–170 km wide and 1.3 km deep. I am grateful to Professor Hildebrand for his clarifications.

77 An online computer game called *Meteors and Dinosaurs* offers the dinosaurs a chance to vindicate their extinction. 'When meteors [sic] threaten to destroy all of the life on Earth, the dinosaurs . . . evolved laser beams in their eyes, as well as missile launchers!', www.arcadeprehacks.com.

78 S. M. Dambrot, 'Not by Asteroid Alone: Rethinking the Cretaceous Mass Extinction', www.physorg.com, 19 January 2012.

79 Zahnle and Sleep, 'Impacts and the Early Evolution of Life', p. 258.

80 The definition of a 'globally catastrophic impact' is contingent on the amount of ash entering the atmosphere in sufficient quantities to lower global temperatures. A drop in several degrees would be enough to drastically reduce crop yields. D. Morrison, 'The Contemporary Hazard of Comet Impacts', in *Comets and the Origin and Evolution of Life*, pp. 303–4.

81 Thanks to Professor Jay Milosh for helping to clarify and quantify this scenario.

82 Impact-related films became a genre in the 1950s. A short list includes *When Worlds Collide* (1951), *The Day the Sky Exploded* (1958), *The Green Slime* (1968), *Meteor* (1979), *Deep Impact* (1998) and the author's personal favourite, *Armageddon* (1998). The impending release of the latter two prompted a u.s. Congressional hearing on space defense, in anticipation of the spike in public concern. M.E.B. France, 'Planetary Defense: Eliminating the Giggle Factor', published by National Defense University, National War College (Washington, DC, 2000), p. 10.

83 From an EBSCOhost (www.ebscohost.com) search for papers on 'mass extinction events'.

84 See France, 'Planetary Defense: Eliminating the Giggle Factor' for a synopsis of u.s.-initiated planetary defence efforts.

85 Susan Sontag, *Against Interpretation* (New York, 1966), p. 42.

86 Gounelle, 'The Meteorite Fall at L'Aigle and the Biot Report', p. 77.

87 Camille Flammarion, *Astronomy for Amateurs*, trans. Francis A. Welby (London, 1903), p. 203.

88 The oft-quoted phrase 'cosmic shooting gallery' seems to have been coined by William J. Broad, 'Earth is Target for Space Rocks at Higher Rate than Thought', *New York Times*, 7 January 1997.

89 NASA's early warning system was named after the one envisaged by Arthur C. Clarke in *Rendezvous with Rama* (New York, 1972).

90 D. A. Crawford, 'Comet Shoemaker-Levy 9, Fragment Size and Mass Estimates from Light Flux Observations', 1997 NASA 28th Annual

Lunar and Planetary Science Conference Paper, available at http://adsabs.harvard.edu.

91 Bryson, *A Short History of Nearly Everything*, p. 243.

92 While NASA predicted the asteroid passage of 15 February 2013, the fireball that exploded over Chelyabinsk, Russia, that same day was unrelated and unexpected.

93 Gail Collins, 'Rocks in Space', *New York Times*, 21 April 2013. The funds so far dedicated to asteroid tracking have been estimated as less than the production budget for Hollywood's *Armageddon*.

94 NASA's NEO count is regularly updated online: http://jpl.nasa.gov/asteroidwatch.

95 Zahnle and Sleep, 'Impacts and the Early Evolution of Life', pp. 303–4.

96 Nield, *Incoming*, p. 180.

97 Except for *An Inconvenient Truth* (2006), the documentary following then u.s. vice president Al Gore, eco-disaster documentaries, however stunning (like *Darwin's Nightmare*, 2004, about the catastrophic consequences of introducing Nile perch to Lake Victoria), cannot match the box office draw of fictional eco-disaster films such as *The Day After Tomorrow* (2004), *The Happening* and *WALL-E* (both 2008), *2012*, *Knowing* and *The Road* (all 2009).

98 V. I. Vernadsky, 'Some Words about the Noösphere' (1943), first published in abridged form in *American Scientist* in 1945 and reprinted in *21st Century Science and Technology*, XVIII/1 (Spring 2005), pp. 16–21, and available at www.21stcenturysciencetech.com.

99 Ibid., p. 19. Vernadsky's three stages of Earth's development are the geophere (prebiotic), the biosphere (biological life) and the noösphere (conscious life). Philosopher Pierre Teilhard de Chardin (1881–1955) referred to 'the sphere of thought' and interaction of human minds in *Cosmogenesis* (1922). Vernadsky attributed the concept of noösphere to philosopher mathematician Edouard Le Roy (1870–1954).

100 John Polk Allen, 'The Evolution of Humanity, Past, Present And Future: A Review of Humanity's Taxonomic Classification and Proposal to Classify Humanity as a Sixth Kingdom, Symbolia', www.biospherics.org (2000), p. 18.

101 Vernadsky, 'Some Words about the Noösphere', p. 20; Prof. Evgenii Shepelev, in his preface to *The Biosphere* (Santa Fe, NM, 1986), p. 2.

102 Nield, *Incoming*, p. 247.

2 Fallen Gods

1 Dr Tony Phillips, 'What Exploded over Russia?', NASA Science News website, www.science.nasa.gov/science-news, 26 February

2013. The article features an infrasound recording of the event, heard up to 15,000 km away.

2 Successive sonic booms signalled the masses' further fragmentation; the largest meteorite punched a hole through the frozen Lake Chebarkul and thousands of smaller ones were scattered in the vicinity. Jiří Borovička et al., 'The Trajectory, Structure and Origin of the Chelyabinsk Asteroidal Impactor', *Nature*, 503 (14 November 2013), pp. 235–7.

3 '"Shock and Frustration": Locals Report on Meteorite Crash in Russian Urals', RT News, www.rt.com, 15 February 2013.

4 A. Kuzmin, 'Russia Cleans Up after Meteor Blast Injures More than 1,000', www.reuters.com, 16 February 2013; 'Meteorite Hits Russian Urals: Fireball Explosion Wreaks Havoc, up to 1,200 Injured', RT News, www.rt.com, 15 February 2013.

5 Viewed 4.6 million times within weeks of its posting: 'Meteorite Explosion – Russia Chelyabinsk 2/15/2013 – Asteroid Expected to Pass Close by Earth', www.youtube.com, 15 February 2013

6 *Harper's Weekly*, 19 February 2013.

7 'Like a Rock: Church of Meteorite Set Up by Worshippers of Famous Space Debris', RT News , www.rt.com, 17 September 2013. This response bears a striking similarity to a fictional one conceived by Vladimir Sorokin in his *Ice Trilogy* (2006). See chapter Four of the present work.

8 Sir James Fraser, *The Golden Bough: A Study in Magic and Religion* (Ware, Hertfordshire, 1993), p. 279. Meteors figured in more prosaic rituals, such as that prescribed as a cure for pimples by Marcellus of Bordeaux, who served in the court of Emperor Theodosius I (*r.* AD 347–95) 'While the star is still shooting from the sky, wipe the pimples with a cloth . . . Just as the star falls from the sky, so the pimple will fall from your body, only be very careful not to wipe them with your bare hand, or the pimples will be transferred to it', ibid., p. 17.

9 *Aeneid*, Book 2, available at www.online-literature.com.

10 Livy notes numerous 'showers of stones' following which 'nine days of rites were observed' a standard Roman practice from the seventh century BC. Alastair MacBeath and Andree Dorian Gheorghe, 'Meteor Beliefs Project: Meteorite Worship in the Ancient Greek and Roman Worlds', *WGN, the Journal of the International Meteor Organization*, XXXIII/5 (2005), p. 139.

11 Camille Flammarion, trans. Francis A. Welby, *Astronomy for Amateurs* (London, 1903), p. 185.

12 J. K. Bjorkman, 'Meteors and Meteorites in the Ancient Near East', *Meteoritics*, VIII/2 (1973), p. 92.

13 Hossein Alizadeh Gharib, 'The Magi and the Meteorites', *Griffith Observer*, September 2002, available at the Cambridge Conference

Network, http://abob.libs.uga.edu/bobk/cccmenu.html, accessed
2 April 2015. Ghrarib quotes the Pseudo Clementine Recognitions
(second century AD), Dio Chrysostum (first century) and
al Maghdassi.

14 Ibid. Similarly, Upper Egyptian folklore holds that meteorites are
the weapons god uses to cast down demons attempting to assault
heaven (conversation with Tarek Salem, Medinat Habu, 9 December
2013). See also MacBeath and Gheorghe, 'Meteor Beliefs Project:
Meteorite Worship in the Ancient Greek and Roman Worlds', p. 135,
for an account of the *ancile*, a magical shield said to have fallen from
heaven amid thunderous blasts during the rule of Rome's second
king, Numa Pompilius (715–673 BC).

15 Ernst Cassirer, *Language and Myth* (New York, 1946), p. 32.

16 Ibid., p. 35.

17 Martin Heidegger, *What is Philosophy?*, trans. William Kluback
(Baltimore, MD, 1956), p. 81.

18 John G. Burke, *Cosmic Debris: Meteorites in History* (Berkeley, CA,
1986), p. 215.

19 Edmund S. Meltzer, 'Indirect Evidence for the Identification of
the Benben as a Meteorite?', *Discussions in Egyptology*, LII (2002),
pp. 81–3.

20 Bjorkman, 'Meteors and Meteorites in the Ancient Near East',
p. 110. The Egyptian–Hittite conflict lasted until 1258 BC.

21 Massimo D'Orazio, 'Meteorite Records in the Ancient Greek and
Latin Literature', *Myth and Geology*, ed. L. Piccardi and W. B.
Masse, Geological Society, London, Special Publications 273 (2007),
p. 219. Only Eusebius has transmitted fragments attributed to the
Phoenician Sanchouniathon, who modern historians consider a
legendary figure dating to the pre-Homeric era.

22 McBeath and Gheorghe, 'Meteorite Worship in the Ancient Greek
and Roman Worlds', p. 138. 'Idaean' derives from the ancient name
for a mountain in Anatolia.

23 Ibid., pp. 138–9.

24 E. M. Antoniadi, 'On Ancient Meteorites, and the Origin of the
Crescent and Star Emblem', *Journal of the Royal Astronomical Society
of Canada*, XXXIII/5 (May–June 1939), pp. 177–8, provided online
by the NASA Astrophysics System, http://adsabs.harvard.edu.

25 Ibid., pp. 180–81.

26 Ibid., pp. 181–2. Antonin Artaud reimagines Elagabalus' meteorite
worship in *Héliogabale on l'anarchiste couronné* (Paris, 1934).

27 Antoniadi, 'On Ancient Meteorites', p. 181. Lampridius was
purportedly one of several authors of *The Augustan History* (covering
AD 117–284), a book whose veracity and authorship remains
controversial.

28 Burke, *Cosmic Debris*, p. 226.
29 Lincoln LaPaz, *Topics in Meteoritics* (Albuquerque, NM, 1969),
    pp. 89–91.
30 Pottery from the burial dates to *c.* 1100–1200. Alastair MacBeath,
    'Meteorite Veneration in the New World', *WGN, the Journal of the
    International Meteor Organization* (December 2010), p. 194.
31 Ibid.
32 Burke, *Cosmic Debris*, p. 224.
33 W. D. Boutwell, 'Tomb of a Giant Meteorite', *National Geographic*,
    LIII (1928), p. 729.
34 Burke, *Cosmic Debris*, p. 224, and Clackamas County, Oregon,
    website: www.usgennet.org.
35 Steve Wilson, *Oklahoma Treasures and Treasure Tales* (Norman, OK,
    1989), p. 87.
36 Christopher E. Spratt, 'Canada's Iron Creek Meteorite', *Journal of the
    Royal Astronomical Society of Canada*, LXXXIII/2 (1989), pp. 81–91.
37 J. Gerson, 'First Nations College Calls for a Return of Sacred
    Meteorite from Alberta Museum', *National Post* (Canada),
    8 August 2012.
38 Burke, *Cosmic Debris*, p. 226.
39 Liliana Samuel, 'Plan to Move Argentine Meteorite to Germany
    Blocked', AFP News, 2 February 2012.
40 Duane W. Hamacher and Ray P. Norris, 'Australian Aboriginal
    Geomythology: Eyewitness Accounts of Cosmic Impacts?',
    Archaeoastronomy, Journal of Astronomy in Culture, pre-print
    submitted 22 September 2010, p. 14. Available online from
    Cornell University: http://arxiv.org.
41 Ibid., p. 9.
42 Jane Gordon quoted in Peggy Reeves Sanday, *Aboriginal Paintings
    of the Wolfe Creek Crater: Track of the Rainbow Serpent* (Philadelphia,
    PA, 2007), p. 103.
43 Stan Brumby quoted ibid., p. 58.
44 Speiler Sturt, quoted in Peggy Reeves Sanday, *Aboriginal Paintings
    of the Wolfe Creek Crater: Track of the Rainbow Serpent* (Philadelphia,
    PA, 2007), p. 15.
45 Ibid., p. 8.
46 Ibid., p. 18.
47 Duane W. Hamacher and Ray P. Norris, 'Meteors in Australian
    Aboriginal Dreamings', *WGN, the Journal of the International Meteor
    Organization*, XXXVIII/3(24 February 2010), p. 91. As in European
    folklore, the Arunta of Australia's Central Desert associate
    mushrooms and toadstools with falling stars.
48 Sven Ouzman, 'Flashes of Brilliance: San Rock Paintings of
    Heaven's Things', in *Seeing and Knowing: Rock Art with and without*

Ethnography, festschrift dedicated to David Lewis-Williams
(Johannesburg, 2010), pp. 11–35. See V. Clube and B. Napier,
The Cosmic Serpent (London, 1982).

49 See Alistair McBeath, 'Meteor Beliefs Project: An Introduction to
the Meteor-dragons Special', *WGN, the Journal of the International
Meteor Organization*, XXXI/6 (2003), pp. 189–91.

50 W. Bruce Masse and Michael J. Masse, 'Myth and Catastrophic
Reality: Using Myth to Identify Cosmic Impacts and Massive
Plinian Eruptions in Holocene South America', *Myth and Geology*,
ed. L. Piccardi and W. B. Masse, Geological Society, London,
Special Publications 273 (2007), p. 193.

51 G.J.H. McCall, A. J. Bowden and R. J. Howarth, 'The History of
Meteoritics – An Overview', *The History of Meteoritics and Key
Meteorite collections: Fireballs, Falls and Finds*, ed. G.J.H. McCall,
A. J. Bowden and R. J. Howarth, Geological Society, London,
Special Publications 256 (2006), p. 4. Similarly, chapels were
erected above the sites of meteorite falls, which were regarded
as evil omens in parts of medieval Russia.

52 Qur'an, Surat Al-Ahqaf, 46:22–6.

53 Philby's adventure is recounted by Elizabeth Monroe, 'Across the
Rub al-Khali', *Saudi Aramco World*, XXIV/66 (November/December
1973), pp. 6–13; 'Desert Meteorites', *Aramco World*, XIII/6 (June–
July 1961), pp. 18–20; and Zayn Bilkadi, 'The Wabar Meteorite',
Aramco World, XXXVII/6 (November/December 1986), pp. 26–30.
All Philby quotes are from H. St John Philby, 'Rub' al Khali: An
Account of Exploration in the Great Desert of Arabia under the
Auspices and Patronage of His Majesty 'Abdul 'Aziz ibn Sa'ud,
King of the Hejaz and Nejd and its Dependencies', *Geographical
Journal*, LXXXI (January 1933), pp. 1–26.

54 'Desert Meteorites'. Geologists working for Aramco Oil identified
six meteorite fall sites in the Empty Quarter as of 1961, in addition
to Wabar.

55 Z. Bilkadi, 'The Wabar Meteorite'; Thomas Abercrombie, 'Saudi
Arabia', *National Geographic*, CXXIX/1 (January 1966), pp. 1–52.

56 In 1992 the Egyptian geologist Farouk el-Baz discovered the actual
remains of Ubar in Oman, 400 km south of Philby's craters, using
satellite imagery.

57 Jeffrey C. Wynn and Eugene M. Shoemaker, 'The Day the Sands
Caught Fire', *Scientific American*, CCLXXIX/5 (November 1998),
pp. 36–43.

58 Elsebeth Thomsen, 'New Light on the Origin of the Holy Black
Stone of the Ka'ba', *Meteoritics*, XV/1 (31 March 1980), pp. 87–91.

59 Antoniadi, 'On Ancient Meteorites', p. 183.

60 Burke, *Cosmic Debris*, pp. 222–3.

61 For the Kaaba stamp and a range of others depicting meteorites, see Philip R. Burns's annotated collection: www.pibburns.com, accessed 4 March 2015.

62 Mark Taylor, 'Priceless Tibetan Buddha Statue Looted by Nazis was Carved from Meteorite', *The Guardian* (UK), 28 September 2012.

63 Lizzy Davies, 'Nazi Buddha from Space Might be a Fake', *The Guardian* (UK), 24 October 2012.

64 Collin Barras, 'Buddhist Statue Acquired by Nazis is a Space Rock', www.newscientist.com, 27 September 2012.

3 To Have and To Hold

1 The Antarctic Search for Meteorites Program (ANSMET) has conducted an annual collecting mission since 1976. ALH84001 is named after the abbreviated location of the find (Alan Hills, Antarctica), the year (1984) and order (first of the mission) in which it was found. See David S. McKay et al., 'Search for Past Life on Mars: Possible Relic Biogenic Activity in Martian Meteorite ALH84001', *Science*, CCLXXIII/5277 (16 August 1996), pp. 924–30. A meteorite's martian origin can be determined in several ways, including when the composition of gas bubbles trapped in its silicate inclusions match atmospheric analyses reported by NASA's Viking mission to Mars.

2 A transcript of President Clinton's speech of 7 August 1996 is available at www2.jpl.nasa.gov/snc/clinton.html, accessed 21 November 2014.

3 Arthur Hirsch, 'How Much for a Piece of This Rock? Meteorites: Small Chunks of the Red Planet Turn Collectors Green with Envy', *Baltimore Sun*, 20 November 1996.

4 Kevin Kichinka, *The Art of Collecting Meteorites* (Ashland, OH, 2005), p. 17.

5 Egypt's ruler Muhammad Ali awarded the British an obelisk for militaristic services rendered to repel Bonaparte's occupation attempt (1798–1801). The 200-tonne granite obelisk lay on an Alexandria beach until 1878 (three years before the British occupation) when it was encased in the 'Cleopatra', a specially designed 'iron cocoon' fitted with rudder and sail, towed by steamship to London and raised amid much fanfare by the Thames.

6 John G. Burke, *Cosmic Debris: Meteorites in History* (Berkeley, CA, 1986), p. 178. The Vienna meteorite collection, the world's oldest, may be dated to the tenure of Abbé Andreas Xaver Stütz (1798–1806) as director of the imperial natural history collection, presented to the state by Empress Maria Teresa in 1765. Stütz catalogued the cabinet's seven meteorites, including the Hrashina

(Croatia, 1751), the founding object of the collection.
7 Ibid., p. 203.
8 Ibid., p. 205.
9 Ibid., pp. 204–5.
10 Ibid.
11 Ibid., p. 205.
12 Kichinka, *The Art of Collecting Meteorites*, p. 34.
13 Ibid., p. 3, quoting Russ Kempton of the New England Meteoritical Services.
14 Paul Harris and Jim Tobin, 'Meteorite People', *Meteorite Times Magazine* (January 2003), available at www.meteorite-times.com.
15 Macovich Collection homepage, www.macovich.com.
16 The 112-page catalogue for the Heritage meteorite auction (New York, 14 October 2012), for which Pitt supplied the text and most of the photographs, is available at www.macovich.com/heritage/6089_catalog.pdf.
17 Ibid., pp. 33, 18, 73.
18 Ibid., pp. 22–9.
19 Ibid., pp. 50, 11.
20 Ibid., pp. 66, 47.
21 Ibid., pp. 76, 2, 72, 75.
22 Author's correspondence with Darryl Pitt, 20 January 2014.
23 Dr Agee's correspondence with D. Pitt, 5 April 2011. The solar system's oldest igneous rocks, only twelve angrites (an achondrite or stony meteorite) are known to science.
24 The Nomenclature Committee of the Meteoritical Society, a non-profit scholarly organization founded in 1933 to promote the study of meteorites and space-mission samples, analyses, classifies, names and numbers meteorites from finds and falls.
25 Author's conversation with Darryl Pitt, 23 December 2013.
26 Author's conversation with Dr Brandstätter in Vienna, 28 August 2013.
27 Jennifer Gould, 'To Have and to Hold', *New York Magazine*, 15 June 2013.
28 Kichinka, *The Art of Collecting Meteorites*, p. 46.
29 Author's correspondence with Dr Harvey, 10 November 2013.
30 Carol Mattack, 'First Came the Russian Meteorite, Now the Meteorite Deals', *Bloomberg Businessweek*, 19 February 2013. The site www.avito.ru is the Russian equivalent of Craigslist.
31 Andrew E. Kramer, 'Russians Wade into the Snow to Seek Treasure from the Sky', *New York Times*, 18 Feburary 2013.
32 Ursula B. Marvin, 'Meteorites in History: An Overview from the Renaissance to the 20th Century', *The History of Meteoritics and Key Meteorite Collections: Fireballs, Falls and Finds*, ed. G.J.H. McCall,

A. J. Bowden and R. J. Howarth, Geological Society, London, Special Publications 256 (London, 2006), p. 37.

33 I cannot locate the precise reference for this note taken while researching antiquity hunting some years ago, but an Egyptologist, Salima Ikram, assures me that ancient artefacts were being forged in late antiquity.

34 International Meteorite Collectors Association homepage. Websites offering instructions for authenticating meteorites outside a lab include www.meteoritemarket.com.

35 Curry was sentenced on 19 October 2013 to 500 hours of community service, a U.S.$1,500 fine plus payment of court costs and reimbursement of monies obtained from fraudulent sales. Steve Weishampel, 'Expensive Meteorites are Bogus, AG Says', *Courthouse News Service*, 23 April 2012.

36 Clara Moskowitz, 'Man Accused of Stealing Meteorites in North Carolina', www.space.com, 3 January 2013.

37 Kichinka, *The Art of Collecting Meteorites*, p. 19.

38 Ibn Khaldun's fourteenth-century *Muqadimmah: An Introduction to History* (1377) speaks at length of the 'weak-minded' individuals who prefer hunting treasure to an honest living, pp. 301–4. See Robert Irwin, *Arabian Nights: A Companion* (London, 1994), pp. 184–9 for a discussion of 'the seekers' and the role of the occult in treasure hunting. Ahmed Kamil, director of the Cairo Museum, translated and published a French edition of *The Book of Buried Pearls* in 1900. His introduction mentions several treasure-hunting treatises, some dating to the 1300s. Kamil credits *The Book of Buried Pearls* with causing more damage to antiquities, by encouraging treasure hunting and pillaging, 'than war or the [passage of] centuries'. I am grateful to the Institut Français d'Archéologie Orientale for allowing me to peruse the rare copy kept in their Cairo library. On nineteenth-century travellers' appetite for Egyptian antiquities see Brian M. Fagan, *The Rape of the Nile* (Cambridge, 2004), pp. 205–16.

39 Addie Nininger and Dorothy S. Norton, who collaborated with their husbands in a variety of meteorite-related endeavours, qualify as 'meteorite women' and there are no doubt others, including Nininger's daughter, Margaret Huss, who operated the American Meteorite Laboratory with her husband Glenn well into the 1980s. Here 'meteorite men' refers to the rich and famous fellows who make a living hunting meteorites and might have called themselves 'meteorite people', but didn't.

40 Kichinka, *The Art of Collecting Meteorites*, p. 51.

41 H. H. Nininger, *Find a Falling Star* (Forest Dale, VT, 1973), pp. 100–111.

42 Transcript of an interview with Nininger conducted on 28 January
    1976, Northern Arizona University Library archives, call number:
    NAU.OH.28.34. Available at http://archive.library.nau.edu.
43 O. Richard Norton, *Rocks from Space* (Missoula, MT, 1998),
    pp. 278–81.
44 Ibid., p. 282. Nininger was the first to find proof that the mass that
    formed Meteor Crater vaporized on impact: tiny 'spheroids' nickel-
    iron droplets that rained from the sky as it cooled. Norton, *Rocks
    from Space*, p. 284.
45 Ibid., pp. 279–87.
46 Ibid., p. 290.
47 Ibid., pp. 289–90.
48 Ibid., p. 290. Haag was also encouraged to hunt meteorites by James
    Maxime DuPont (1912–1991), who assembled a large personal
    collection he bequeathed to the Planetary Studies Foundation,
    an Illinois non-profit organization dedicated to 'bringing a better
    understanding and appreciation of the universe'.
49 Robert Haag, *The Field Guide of Meteorites* (Tucson, AZ, 1997),
    p. 36. Witnesses of some meteorite falls have reported the smell of
    sulphur, but the Murchison fall's powerful odour is apparently
    unique.
50 Harris and Tobin, 'Meteorite People'. Pitt names Haag and the late
    Dr Marty Prinz of the American Museum of Natural History as
    the people who inspired his career as meteorite collector-dealer.
51 Geoffrey Notkin, *Rock Star: Adventures of a Meteorite Man* (Tucson,
    AZ, 2012), p. 39.
52 Rare earth magnets, aka neodymium magnets, the strongest and most
    affordable permanent magnet, are used in a variety of applications,
    including computer hard drives. Ibid., pp. 2–3.
53 Ibid., p. 129.
54 Ibid., p. 240.
55 Norton, *Rocks from Space*, p. 307.
56 Kichinka, *The Art of Collecting Meteorites*, pp. 30–31.
57 The Meteoritical Society for Meteoritics and Planetary Science:
    www.lpi.usra.edu/meteor/index.php.
58 Jeff Wolf, 'Deiro: The Count, Hughes Pilot', *Las Vegas Review-
    Journal*, 18 September 2008. A star of America's vaudeville circuit
    in the early 1900s, the first Count Diero was briefly married to
    actress Mae West.
59 Paul Harris and Jim Tobin, 'Meteorite People', *Meteorite Times
    Magazine* (March 2010), available at www.meteorite-times.com.
60 Author's correspondence with Sonny Clary, 6 September 2013.
61 Kichinka, *The Art of Collecting Meteorites*, pp. 54–6.
62 Norton, *Rocks from Space*, p. 169.

63 The annual Tucson Gem Show is the meet-and-greet epicentre of international meteorite people, many of whom know and collaborate with each other on various projects.

64 Tom Phillips's online gallery: www.tomphillipsmicro.com.

65 Derek Sears, researcher at NASA Ames Research Center and Robert Beauford, doctoral candidate at the Arkansas Center for Space and Planetary Sciences at the University of Arkansas, co-edit *Meteorite Magazine* (founded in 1955); Paul Harris and Jim Tobin, hunters/collectors since the early 1990s, started Meteorite-times.com in 2002.

66 G. K. Chesterton, *The Club of Queer Trades* (Ware, 1995), p. 2.

4 All Things Said and Done

1 Ursula Marvin, 'The Meteorite of Ensisheim, 1492–1992', *Meteoritics*, XXVII (March 1992), pp. 28–72.

2 Ibid., fig. 1a, p. 30.

3 Ibid., p. 34.

4 Ibid., p. 43.

5 *A Heavenly Body* (1494–6) is kept in the National Gallery, London. Ursula B. Marvin, 'Meteorites in History: An Overview from the Renaissance to the 20th Century', from *The History of Meteoritics and Key Meteorite Collections: Fireballs, Falls and Finds*, ed. G.J.H. McCall, A. J. Bowden and R. J. Howarth, Geological Society, London, Special Publications 256 (London, 2006), fig. 6, p. 23.

6 Ibid., p. 17.

7 Raphael's *Madonna of Foligno* (Vatican Museums, 1511–12) includes a detail of the fireball seen over Crema (70 km from Foligno) on 4 September 1511 beneath an arc of light, when Maximilian, now Holy Roman Emperor, was still battling the French. Placing a fireball in an otherwise highly conventionalized Madonna harkened deliberately to the Ensisheim fall, implying 'divine reconciliation and assistance' once again for Maximilian. See Lincoln laPaz, *Topics in Meteoritics* (Albuquerque, NM, 1969), pp. 94–7.

8 W. G. Guthrie, 'The Astronomy of Shakespeare', *Irish Astronomical Journal*, VI/6 (June 1964), pp. 201–11.

9 Shakespeare saw the passage of (Halley's) Comet in 1607, before Halley had plotted its periodicity.

10 Guthrie, 'The Astronomy of Shakespeare', p. 201. Shakespeare may have heard of Copernican theory via Bruno, who lectured in England for fourteen years before he was burned at the stake for heresy in Rome.

11 Dorothea Havens Chappell, 'Shakespeare's Astronomy', *Publications of the Astronomical Society of the Pacific*, LVII/338 (October 1945), p. 256.

12 Ibid., p. 257.

13 Ibid., p. 255.

14 Ibid., p. 259.

15 Lori Stiles, 'Eugene Shoemaker Ashes Carried on Lunar Prospector', *University of Arizona News Services*, 6 January 1998, www2.jpl.nasa.gov. I can find no other record of Shakespeare's works being sent to space, though a page from his first folio was considered for the 'golden record' mounted on the *Voyager 1* and *Voyager 2* probes launched in 1977. Instead, an analogue-encoded page from Isaac Newton's 'system of the world' was chosen, where the process of launching an object into space was first essayed. Megan Gambino, 'What is on Voyager's Golden Record', www.smithsonian.com, 23 April 2012.

16 See chapter One of the the present work for Shoemaker's role in the history of meteoritics. The capsule with his ashes was carried on NASA's *Lunar Prospector*, which crash-landed as programmed on the moon on 31 July 1999.

17 John Donne, 'An Anatomy of the World', quoted in Thomas S. Kuhn, *The Copernican Revolution: Planetary Astronomy in the Development of Western Thought* (Cambridge, 1957), p. 194.

18 Donne, 'The First Anniversarie', in Alistair McBeath and Andrei Dorian Gheorghe, 'Meteor Beliefs Project: Meteors in the Poems of John Donne', *WGN, the Journal of the International Meteor Organization*, XXXII/3 (2004), pp. 92–4.

19 *A Hundred Wonders of the Modern World and of the Three Kingdoms of Nature* was published by Phillips (1767–1840) under the pseudonym Reverend C. C. Clarke and was illustrated with numerous engravings.

20 See chapter One of present work.

21 Austin's short story appeared in *The Token and Atlantic Souvenir* (1839), an illustrated annual collection of stories and poems.

22 Poe's story was published in *Burton's Gentlemen's Magazine*. Quoted excerpts from Thomas Olive Mabbott, ed., *The Collected Works of Edgar Allan Poe*, vol. II: *Tales and Sketches* (Cambridge, 1978), pp. 451–62.

23 Ibid.

24 Ibid.

25 Ibid., p. 455.

26 William S. Burroughs, *Cities of the Red Night* (London, 1981), p. 155. Burroughs shared Poe's penchant for plague and catastrophe. As part of the soundtrack of a 1995 computer game, *The Dark Eye*, based on Poe's oeuvre, Burroughs reads 'The Masque of the Red Death' (1842) and Poe's last poem, 'Annabelle Lee', with ghastly gusto (videos are available on YouTube).

27 William Kelly Simpson, ed., *The Literature of Ancient Egypt* (Cairo, 2003), p. 51. For a discussion of a theoretically catastrophic impact

in ancient Egypt, *see* Aly Barakat, 'Did the Kamil Meteorite Fall Contribute to the Downfall of the Old Kingdom?', *The Ostracon: Journal of the Egyptian Study Society*, XXIV (Fall 2013), pp. 12–21.

28 The fireball of 1860 (and another depicted by Gustave Hahn in 1913 (see image p. 141) are sometimes referred to as 'earth-grazers', meteoroids that skip over earth's atmosphere like a stone on a pond, and depart intact without depositing meteorites. But according to Dr Pavel Spurný, head of the Ondřejov Observatory's Department of Interplanetary Matter, both Church and Hahn illustrate the fireballs' 'substantial fragmentation with noticeable deceleration (smaller pieces are more decelerated) [therefore] it is not possible that part of such a body could survive the atmospheric flight and escape the atmosphere and Earth's gravity'. Spurný interprets both 'very realistic paintings as clear examples of the terminal part of the luminous flight of a big bolide which produced a standard multiple meteorite fall.' (Author's correspondence, 11 September 2013.)

29 H. G. Wells, *The War of the Worlds* (New York, 2002), p. 11.

30 Ibid., pp. 11–12.

31 Ibid., p. 3.

32 Jules Verne, *The Chase of the Golden Meteor* (Lincoln, NE, 1998), p. 36.

33 Ibid., pp. 65, 85.

34 Addressing a broader, inquiring readership, French astronomer Camille Flammarion's (1842–1925) *Astronomy for Women* (1901) opened with women's contributions to the field. A 1903 edition of the English translation by Frances A. Welby was re-titled *Astronomy for Amateurs*.

35 Doyle's *The Valley of Fear* was published in instalments by the *Strand Magazine* in 1914–15.

36 Verne, *The Chase of the Golden Meteor*, quoted in the introduction by Gregory A. Benford, p. ix.

37 Antoine de Saint-Exupéry, *Wind, Sand and Stars*, trans. Lewis Galantière (New York, 2002), pp. 62–4.

38 O. Richard Norton, *Rocks from Space* (Missoula, MT, 1998), p. 89.

39 Ibid., p. 91.

40 Ibid.

41 For Barringer's endeavour, see chapter One of the present work.

42 Vladimir Sorokin, trans. Jamey Gambrell, *Ice Trilogy* (New York, 2008), pp. 4, 34.

43 Ibid., pp. 38–41.

44 Norton, *Rocks from Space*, pp. 92–6.

45 Ibid., p. 97. In the late 1950s analyses of Kulik's soil samples yielded particles of meteoritic dust.

46 Sorokin, *Ice*, p. 44.

47 Ibid., p. 48.
48 Ibid., pp. 57–60.
49 Ibid., p. 62.
50 Ibid., back cover, by Ingo Schulze.
51 Ibid., p. 217.
52 Anton Yevseev, 'Mystery of Tunguska Meteorite Solved', 25 October 2010, *Pravda* English online, www.pravda.ru.
53 Thomas Pynchon, *Against the Day* (New York, 2006), p. 5.
54 Quoted in the superb introduction by Kitty Mrosovsky to her translation of Flaubert's *The Temptation of Saint Anthony* (Harmondsworth, 1980), p. 11.
55 Pynchon, *Against the Day*, p. 782.
56 Ibid., p. 784.
57 Ibid., p. 785.
58 A 2009 model created by a private American firm, Risk Management Solutions (RMS), catering to 'insurers, reinsurers, trading companies, and other financial institutions', showed that had a Tunguska-sized event occurred over lower Manhattan, it would have killed half the inhabitants and levelled 70 per cent of everything from Queens and the Bronx to north New Jersey.
59 Norton, *Rocks from Space*, pp. 102–5.
60 Ian Sample, 'Scientists Reveal the Full Power of the Chelyabinsk Meteor Explosion', *The Guardian* (UK), 7 November 2013.
61 Alice Bona, 'Everybody Must Get Stoned', interview with Maurizio Cattelan, for www.christies.com, 4 November 2001, in anticipation of an auction on 17 May in which *The Ninth Hour* sold for U.S.$886,000.
62 Parker is quoted in a piece describing the work on the British Council collection website, 'Cornelia Parker (1956–)', http://visualarts.britishcouncil.org.
63 Author's correspondence with Regine Petersen, 3 February 2014.
64 Coline Milliard, 'Staring into Space: Katie Paterson Shoots for the Heavens', *Modern Painters* (June 2012), pp. 56–9.
65 Matthew Rose, 'Letter from Berlin – Forgeries, Pheromones and Clones, Ten Questions for Jonathon Keats', 13 May 2012, www.theartblog.org. Similarly, a Chilean cabernet ('Meteorito') was aged in an oak barrel with a meteorite. 'When you drink this wine, you are drinking elements from the birth of the solar system', says vintner and astronomer Ian Hutcheon. Tara Kelly, 'Meteorite Wine Unveiled in Chile', *Huffington Post*, www.huffingtonpost.com, 24 January 2012.
66 Stan Gaz, *Sites of Impact* (Princeton, NJ, 2009), quote from Robert Silberman's engaging essay 'Between Heaven and Earth: The Impact Photographs of Stan Gaz', p. 27.

5 Strange Landings

1 The gem was once thought to be quartz. 'Tut's Gem Hints at Space Impact', BBC News online, www.bbc.co.uk/news, 19 July 2006.

2 I have Darryl Pitt to thank for this list of Texas meteorites.

3 Lincoln LaPaz, *Topics in Meteoritics* (Albuquerque, NM, 1969), p. 87.

4 Marvin, 'Meteorites in History', p. 25.

5 Ibid., p. 24.

6 Bradley E. Schaefer, 'Meteors that Changed the World', *Sky and Telescope Magazine*, 1 February 2005.

7 D. Johnson, M. M. Grady and J. Tyldesley, 'Gerzeh, a Prehistoric Egyptian Meteorite', *Meteoritics and Planetary Science*, 46(s1) (2011), pp. A114–A114.

8 J. G. Burke, *Cosmic Debris: Meteorites in History* (Berkeley, CA, 1986), p. 232. On display at the Freer Gallery of Art, Smithsonian Institution, Washington, DC.

9 Alastair McBeath, 'Meteor Beliefs Project: Meteorite Veneration in the New World', *WGN, the Journal of the International Meteorite Association*, XXXVIII/6 (2010), p. 196. The Hopewell people traded meteorite as a precious metal, as did the Inuit inhabiting Northern Greenland from *c.* AD 1000.

10 'Winners at Sochi Winter Olympics to Receive Pieces of Russia Meteorite', *The Telegraph* (UK), 26 July 2013.

11 Meredith Bennett-Smith, '4-Billion-year-old, Extremely Valuable Meteorite Used as Doorstop by Tennessee Family', *Huffington Post*, 22 October 2012.

12 Burke, *Cosmic Debris*, p. 223.

13 The Peekskill Car, presently owned by the Macovich Collection, has its own website: www.meteoritecar.com.

14 *Civilization* has been rated one of the 'top games of all time', along with *Super Mario Brothers* and *Tetris*, www.empireonline.com/features/100greatestgames.

15 *Prayers of Steel* (1920).

16 Ted Nield, *Incoming!* (London, 2011), p. 24.

17 Peter Jenniskens et al., 'The Mbale Meteorite Shower', *Meteoritics*, XXIX/2 (March 1994), pp. 246–54.

18 'Meteor Girl One in a Billion', BBC news online, www.bbc.co.uk/news, 27 August 2002.

19 Kevin Yau, Paul Weissman and Donald Yeomans, 'Meteorite Falls in China and some Related Human Casualty Events', *Meteoritics*, XXIX/6 (November 1994), pp. 864–71.

20 David Spiegelhater, 'Afraid of Being Hit by a Meteorite?', *The Guardian* (UK), 13 October 2011.

21 David Cross, 'What are the Odds Your House will be Destroyed by

an Asteroid?', www.movoto.com, accessed 2 October 2014.

22 Kim Willsher, 'Comette Family Home Damaged by Egg-sized Meteorite', *The Guardian* (UK), 10 October 2011.

23 Camille Flammarion, *Astronomy for Amateurs*, trans. Frances A. Welby (London, 1903), p. 211.

24 Zdeněk Ceplecha, 'Preliminary Notes on Some Results of Photographic Multiple Meteorite Fall of Přibram', *Smithsonian Contributions to Astrophysics*, VII (1965), p. 195.

25 For up to date reports of meteor fireball sightings culled from a variety of sources http://lunarmeteoritehunters.blogspot.com.

26 Author's conversation with Pavel Spurný, Ondřejov, 10 September 2013.

27 Pavel Spurný et al., 'The Bunburra Rockhole Meteorite Fall in SW Australia: Fireball Trajectory, Luminosity, Dynamics, Orbit, and Impact Position from Photographic and Photoelectric Records', *Meteoritics and Planetary Science*, XLVII/2 (2012), pp. 163–85.

28 Pavel Spurný, Jakub Haloda and Jiři Borovicka, 'Benešov Bolide – Surprising Outcome of an Exceptional Story after Twenty Years', proceedings of the International Meteor Conference, 2012. Full publication forthcoming.

29 Martin Horejsi, 'Friendly Fire from Space: Berlanguillas, Spain', www.meteorite-times.com, 1 January 2012.

30 At www.factum-arte.com. While only a 'superficial likeness' was requested in the case of the meteorite, Tut's tomb (installed beside the Theban necropolis in 2014) and the Veronese painting are considered the most sophisticated replicas ever fabricated.

31 Paul Strauss, 'Chocolate Meteorites Fill that Crater in your Stomach', 10 April 2013, www.technabob.com. L'eclat also produced a chocolate solar system with bite-size planets including Mercury (coconut mango), Venus (cream lemon) and Earth (cacao).

32 Dawn Hodson, 'Rafferty's Candies Celebrates its One-year Anniversary' *Mountain Democrat* (Placerville California), 31 January 2014.

33 Peter Pham, 'Dogfish Releases Beer Made with Actual Meteorites', 2 October 2013, www.foodbeast.com.

34 Guerlain website (www.guerlain.com) and online review at www.thescentsofself.com, 19 December 2010.

35 Descriptive copy on Amazon.com, accessed 7 February 2015.

SELECT BIBLIOGRAPHY

Branstätter, F., L. Ferrière and C. Köberl, *Meteorites, Witnesses of the Origin of the Solar System* (Vienna, 2013)

Burke, J. G., *Cosmic Debris, Meteorites in History* (Berkeley, CA, 1986)

Gaz, Stan, *Sites of Impact: Meteorite Craters around the World*, photographs by Stan Gaz, essays by Christian Köberl and Robert Silberman (Princeton, NJ, 2009)

Haag, Robert, *A Field Guide to Meteorites* (Tucson, AZ, 1997)

Kichinka, Kevin, *The Art of Collecting Meteorites* (Tucson, AZ, 2005)

McCall, G.J.H., A. J. Bowden and R. J. Howarth, eds, *The History of Meteoritics and Key Meteorite Collections: Fireballs, Falls and Finds*, Geological Society, London, Special Publication 256 (2006)

Marvin, Ursula B., 'Ernst Florens Friedrich Chladni (1756–1827) and the Origins of Modern Meteorite Research', *Meteoritics and Planetary Science*, 31 (1996), pp. 545–88

Nield, Ted, *Incoming! or, Why we Should Stop Worrying and Learn to Love the Meteorite* (London, 2011)

Nininger, H. H., *Find a Falling Star* (Forest Dale, VT, 1973)

Norton, O. Richard, *Cambridge Encyclopedia of Meteorites* (Cambridge, 2002)

—, *Rocks From Space* (Missoula, MT, 1998)

Notkin, Geoffrey, *Rock Star: Adventures of a Meteorite Man* (Tucson, AZ, 2012)

Piccardi, L., and W. B. Masse, eds, *Myth and Geology*, Geological Society, London, Special Publication 273 (2007)

Sanday, Peggy Reeves, *Aboriginal Paintings of the Wolfe Creek Crater: Track of the Rainbow Serpent* (Philadelphia, PA, 2007)

Sorokin, Vladimir, *Ice Trilogy*, trans. Jamey Gambrell (New York, 2008)

Thomas, Paul J., et al., eds, *Comets and the Origin and Evolution of Life* (Berlin, Heidelberg and New York, 2006)

ASSOCIATIONS AND WEBSITES

Aerolite Meteorites
www.aerolite.org
Site of meteorite hunter Geoffrey Notkin's company, Aerolite Meteorites, with hunting and buying information.

ANSMET, The Antarctic Search for Meteorites
http://artscilabs.case.edu/ansmet
U.S.-led field-based science project that recovers meteorite specimens from Antarctica. Funding for annual fieldwork is supported by competed grants awarded to Case Western Reserve University from NASA, while curation and characterization work is supported by a partnership between NASA and the Smithsonian Institution.

Asteroid Watch
www.jpl.nasa.gov/asteroidwatch
Run by NASA/the Jet Propulsion Laboratory, California Institute of Technology. Aims to detect, track and characterize potentially hazardous asteroids and comets that could approach the Earth.

The British and Irish Meteorite Society (BIMS)
www.bimsociety.org
Society for meteorite study and collecting, formed of meteorite researchers and amateur collectors.

Center for Meteorite Studies
http://meteorites.asu.edu
Established in 1961 at Arizona State University following the purchase of a significant portion of H. H. Nininger's collection, one of the world's largest at that time. In addition to curating and augmenting the collection, the centre performs research and offers educational programmes.

International Meteor Organization (IMO)

www.imo.net
International cooperation network of aficionados and amateur astronomers contributing to the study of meteoritic phenomenon. Founded in 1988, it sponsors the Meteor Beliefs Project, a pool of knowledge regarding the beliefs and practices associated with meteoritic phenomena throughout history.

Largest Meteorite Collections in the World

www.jensenmeteorites.com/meteorite-collections.htm
A list of the 22 largest meteorite collections, held at museums and institutions around the world. Includes links to online catalogues.

Latest Worldwide Meteor/Meteorite News

www.lunarmeteoritehunters.blogspot.com
Up-to-date reports of meteor sightings and meteorite falls culled from a variety of sources.

The Macovich Collection of Meteorites

www.facebook.com/MacovichCollection
Extensive aesthetic meteorite collection curated and presented by Darryl Pitt.

Meteorite

www.meteoritemag.org
International quarterly magazine for meteorites and meteorite science.

Meteorite Bulletin Database

www.lpi.usra.edu/meteor/metbull.php
Searchable database of *Meteorite Bulletin*, published by The Meteoritical Society and the primary source for new meteorite-related information.

The Meteorite Exchange

www.meteorite.com
Offers information about buying and selling meteorites in addition to a classification table outlining the properties and names of meteorite types.

The Meteorite Market

www.meteoritemarket.com
Tips for tentatively authenticating meteorites outside a lab.

Meteorite Times

www.meteorite-times.com
Online magazine *Meteorite Times*.

The Meteoritical Society
www.meteoriticalsociety.org
Non-profit scholarly organization founded in 1933 to promote
research and education in planetary science with emphasis on studies
of meteorites and other extraterrestrial materials, including samples
from space missions. The Nomenclature Committee of the Meteoritical
Society is responsible for the naming of all official meteorites.

Near Earth Object Program
http://neo.jpl.nasa.gov/faq
NASA's Near-Earth Object (NEO) tracking program; the NEO count
is regularly updated online.

Tom Phillips Thin Sections – Meteorite and Rock Art
www.tomphillipsrockart.blogspot.com
Gallery of Tom Phillips's micrographs of meteorite thin sections.

The Tricottet Collection
www.thetricottetcollection.com
Collection of rare and aesthetic specimens, an extensive Library of
Meteoritics and an archive of original manuscripts, correspondence
and other rare memorabilia.

ACKNOWLEDGEMENTS

This book began following a conversation with my old friend Max Rodenbeck, long-time chief Middle East correspondent for *The Economist*, about things we'd love to write about besides Egypt, our home for many years. Meteorites had captured my imagination in 1999, when I heard of their role in determining the biochemical origins of life on Earth during a talk given by Chris McKay of the Ames Research Center at an Institute of Ecotechnics conference. As a fellow of the Institute, I attended many of these conferences in Aix-en-Provence that united scientists, artists and thinkers around themes in the cognitive and planetary sciences while highlighting advances in varied technologies. Founded in 1973 by a group of like-minded individuals who had met during the 1960s heyday of Haight-Ashbury, the Institute's goal was to develop a discipline that would harmonize ecology and technology, forces increasingly at odds.

Their grandest effort was Biosphere 2, built in Oracle, Arizona, a miniature Earth enclosed in glass (the size of two football fields) replete with 3,800 living species – everything from microorganisms to humans. Eight people lived inside it for two years, partaking in a unique experiment whose scale, daring and hard data have only now begun to be properly assessed. The purpose was to observe how humans interact with each other and their environment under specific limitations, and how much air, soil and water they need to thrive, while envisaging how, if humanity is to travel into distant space, we might best manage the long journey.

The project was launched with a conference in the former Motorola research centre in Oracle in 1985, the glasnost era. Among those present, aside from luminaries of the Royal Society, NASA and the Smithsonian, were the valiant Russian scientists who first experimented with closed systems, contributed to building the Mir space station and sent plants and small animals up with the cosmonauts, dreaming of an interplanetary future for Earth's life forms. In the captivating atmosphere of that large timbered room, I apperceived the epic dimension of humanity's quest for knowledge, and was gratefully struck by it once more while perusing the extraordinarily wide-ranging research concerning meteorites. This book is intended as a homage to the scientists, scholars and aficionados who have contributed to that remarkable body of work.

NASA's Astrophysics Data System, a vast online archive, was an invaluable research tool, as were the scholarly papers produced by the Meteor Beliefs Project (sponsored by the International Meteor Organization). For generously offering time, knowledge and images, I thank Pavel Spurný of the European Fireball Network and Ondřejov Observatory; Dr Mikhail Nazarov at Moscow's Vernadsky Institute of Geochemistry; Ralph Harvey, chief geologist for the Antarctic Search for Meteorites Program; Franz Brandstätter and Ludovic Ferriere, co-curators of the Vienna Museum of Natural History meteorite cabinet; and Dr Aly Barakat of Egypt's Geological Survey.

I had the pleasure of corresponding with Darryl Pitt, curator of the Macovich Collection, who shared his experience and superb image archive. For contributions of artworks or research-related items I thank meteorite hunter Sonny Clary, Arnaud Mignan (curator of the Tricottet Collection), Sina Najafi of *Cabinet* magazine, Dorothy Norton, Gary Huss, Jonathon Keats, Peggy Schaller, Stan Gaz, Li Wei, Peggy Reeves Sanday, René Paul, Regine Petersen, Alex Kuno, Sven Ouzman, Ben Churcher, Philip R. Burns, Paul Harris, Adam Lowe, Marie-Louise von Wartburg, Lisa Viezbicke, Donald Olson, David Portree, Camille Carlisle, Emma Stratton and Judith Filenbaum Hernstadt, with special thanks to Tom Phillips for his jewel-like micrographs, published here for the first time.

For their help and encouragement I thank Bernd Lötsch, William Lyster, Kathelin Gray, David Golia, Frank Golia, Elizabeth Bolman, Dominic Herbert, Rana Rahimpour, Negar Azimi, Bernard Guillot, Tania Kamal el-Din, Sergei Ivanov, Michael March, Marina Warner, Lara Baladi, Salima Ikram, Professor George Scanlon, David Coulson, John and Clara Semple, Dan Morrison, Pat Lancaster, Younus Porteous, Farouk El-Baz, Vincent Rondot, Amr Zant, Tarek Salem, Rania Awad, Sobhi Saber Hassan, Marie-Jeanne Mozziconacci, Chris Stander, and Pavel and Gabina Sebek, who gave me my first 'meteorite'. I remain indebted to the Institute of Ecotechnics, whose work is present in these pages, as is, somewhere in the background, the music of Jack DeJohnette and Ornette Coleman.

PHOTO ACKNOWLEDGEMENTS

The author and publishers wish to express their thanks to the below sources of illustrative material and/or permission to reproduce it. Locations of some artworks not listed in the captions for reasons of brevity are also given below.

From *Amazing Stories* (August, 1927): p. 134; collection of the author: pp. 7, 111; photo by and reproduced courtesy of Aly Barakat: p. 104; British Museum, London (photo © the Trustees of the British Museum): p. 62 (top); photo © the Trustees of the British Museum, London: p. 129; courtesy of Philip R. Burns: pp. 42, 168; courtesy of the artist (Maurizio Cattelan), Marian Goodman Gallery and Galerie Emmanuel Perrotin, Paris: p. 152; courtesy of Ben Churcher: p. 60; digital image courtesy of Sonny Clary: p. 117; courtesy of Cari Corrigan, ANSMET: p. 103; courtesy of Disney Films: p. 45; The Egyptian Museum, Cairo: p. 162; courtesy of Adam Lowe, Factum Arte: p. 173; courtesy of Ms Judith Filenbaum Hernstadt: p. 132; from Camille Flammarion, *Astronomy for Amateurs* (London, 1903): p. 57; Freer Gallery of Art and Arthur M. Sackler Gallery, Washington, DC: p. 164; photo Friends of Battersea Park: p. 93; Galleria Nazionale d'Arte Antica, Rome – Palazzo Corsini: p. 56; courtesy of Stan Gaz: pp. 36–7, 69, 159; photo Richard Goodbody/Butterfield & Butterfield: p. 99; from Amédée Guillemin, *Le Ciel: notions élémentaires d'astronomie physique* (Paris, 1877), courtesy of Bernd Lötsch: pp. 31, 136; courtesy of Bernard Guillot: p. 77; reproduced courtesy of Judith Filenbaum Hernstadt: p. 132; courtesy Jonathon Keats and Modernism Gallery, San Francisco: pp. 156, 157; courtesy of Alex Kuno: p. 21; photo Iris Lang, courtesy of Darryl Pitt: p. 166; courtesy of Li Wei, www.liweiart.com: p. 160; courtesy of the Library of Congress, Washington, DC (Prints and Photographs Division): pp. 66, 67, 87; courtesy of the Macovich Collection: pp. 38, 89; photo Mark Mauthner/Heritage Auctions: pp. 28, 32, 95, 98, 100, 101; Museo Capitolino, Rome: p. 63; courtesy of NASA/GSFC/Arizona State University: p. 33; courtesy of NASA/JPL/Cornell: p. 10; courtesy of NASA/JPL-Caltech: pp. 47, 48–9, 175; courtesy of NASA/JPL-Caltech/UCLA: p. 8; courtesy of NASA/JPL -Caltech/UCLA/MPS/DLR/IDA: p. 27; courtesy of NASA/JPL-Caltech/University of Arizona: p. 44; Natural History Museum, London (Sowerby Collection) – photo © The Trustees of the Natural History Museum: p. 14 (top);

National Gallery, London: p. 124; Naturhistorisches Museum Vienna collection (photo © Naturhistorisches Museum, Wien/L. Ferrière): p. 82; courtesy of Dr Mikhail A. Nazarov of the Vernadsky Institute, Moscow: pp. 62 (foot), 143, 144, 145, 151; from the *New York Times* (1911): p. 6; photo Addie Nininger, reproduced courtesy of the American Meteorite Laboratory Photo Collections, Collections Research for Museums, Denver, Colorado: p. 109; courtesy of Sven Ouzman: p. 73; courtesy of Regine Petersen: p. 167; courtesy of Tom Phillips: pp. 12, 61, 64, 74, 79, 81, 94, 108, 119, 120, 148–9, 154, 155; courtesy of Darryl Pitt: pp. 88, 90, 91, 123; photo Darryl Pitt/Macovich Collection: pp. 22, 55, 92, 97, 115, 153, 163, 174; photo Hal Povenmire, courtesy of Darryl Pitt: p. 169; from Peggy Reeves Sanday, *Aboriginal Paintings of the Wolfe Creek Crater: Track of the Rainbow Serpent* (Philadelphia, PA, 2007), reproduced courtesy of Peggy Reeves Sanday: pp. 70, 71; from Diebold Schilling's manuscript *Schweizer Bilderchronik des Luzerners* of 1513 at the Zentral- und Hochschulbiblio-thek Luzern: p. 125; courtesy of Pavel Spurný, Ondřejov Observatory, Czech Republic: pp. 9, 170, 172; Tate, London: p. 52; from the Tricottet Collection Image Archive at www.thetricottetcollection.com (courtesy Dr Arnaud Mignan): p. 19; University of Toronto Archives (A2008-0023), © Natalie McMinn: p. 141; from Jules Verne, *La Chasse au météore* (Paris, 1908), courtesy of www.renepaul.net: pp. 137, 138; from Edmund Weiss, *Bilder-Atlas der Sternenwelt: Eine Astronomie für Jedermann . . .* (Esslingen, 1888): p. 29.

INDEX